Laboratory and Scientific Computing

WILEY-INTERSCIENCE SERIES ON LABORATORY AUTOMATION

W. JEFFREY HURST Editor
Hershey, Pennsylvania

ADVISORY BOARD

Tony Beugelsdijk
Los Alamos National Laboratory
Los Alamos, New Mexico

Gary Christian
University of Washington
Seattle, Washington

Ray Dessy
VPI and State University
Blacksburg, Virginia

Tom Isenhour
Duquesne University
Pittsburgh, Pennsylvania

H.M. "Skip" Kingston
Duquesne University
Pittsburgh, Pennsylvania

Gerald Kost
University of California
Davis, California

M.D. Luque de Castro
University of Cordoba
Cordoba, Spain

Frank Settle
VMI Research Institute
Lexington, Virginia

Jerry Workman
Perkin-Elmer
Norwalk, Connecticut

Laboratory and Scientific Computing
A Strategic Approach

JOE LISCOUSKI
Laboratory Automation Standards Foundation
Groton, MA

A Wiley-Interscience Publication
JOHN WILEY & SONS, INC.
New York • Chichester • Brisbane • Toronto • Singapore

This text is printed on acid-free paper.

Copyright © 1995 by John Wiley & Sons, Inc.

All rights reserved. Published simultaneously in Canada.

Reproduction or translation of any part of this work beyond that permitted by Section 107 or 108 of the 1976 United States Copyright Act without the permission of the copyright owner is unlawful. Requests for permission or further information should be addressed to the Permissions Department, John Wiley & Sons, Inc., 605 Third Avenue, New York, NY 10158-0012.

Library of Congress Cataloging in Publication Data:
Liscouski, Joseph G., 1945–
    Laboratory and scientific computing : a strategic approach / by Joe Liscouski.
       p. cm. -- (Wiley-Interscience series on laboratory automation ; v. 1)
    "A Wiley-Interscience publication."
    Includes index.
    ISBN 0-471-59422-9
    1. Laboratories--Data processing. 2. Laboratories--Automation. 3. Science--Data processing. I. Title. II. Series.
Q183.A1L56 1994
542'.85'421--dc20   94-9276

Printed in the United States of America

10 9 8 7 6 5 4 3 2 1

Preface

The early driving factors for laboratory automation were based on the possibility of improving the turn-around time of analysis, the possibility of more accurate and thorough data treatment and improved reporting. Today results, not possibilities, are what count. Yet with all the advances in computer hardware and software, not too much has changed in the application of intelligent devices to the work performed in testing and research facilities. Experimental approaches to data acquisition has been replaced with commercial products that work on a regular basis, but the real gains that were expected are still 'possibilities' except in the few labs that have really made an effort to make things work together.

Much of that effort was chronicled in publications such as *American Laboratory* and *Analytical Chemistry*. In particular is the series of articles by Dr. Ray Dessy in *Analytical Chemistry*, published in the early 1980s and spanning several volumes of that periodical, which described the technologies available for laboratory automation, each technology description being followed by case studies.

Part of the problem is that laboratory automation has been managed and implemented on a 'tactical' basis from it's earliest beginnings—tactical in the sense that projects were limited in scope and planned as isolated implementations that did not consider the impact of that work on the lab's overall computing and information handling needs. Strategic approaches, the primary focus of this book, look at the laboratory as a whole. Each automation project is considered in terms of the immediate problem that has to be solved, and how the data handling needs of that effort mesh with the overall information handling of the entire facility.

The benefits of this method are:

- cost reduction as a result of eliminating faulty design,
- improvements in validation programs and reduced cost of system validation
- improvements in the laboratories ability to meet regulatory requirements
- better use of data and information,
- provision of a design basis for an integrated system,

- consideration of the needs and impact of linking laboratory systems with other systems in an organization, and
- ability to take advantage of new technologies without causing major upheavals in system implementation.

The basic methods used in this book are not limited to laboratory applications. Laboratory automation is an instance of laboratory application. The same approach can be used in process control systems, administrative systems, and other computing applications. One of the strengths of this book is its applicability to other organizations. This means that the information handling requirements of different groups can be integrated though a single set of principles, rather than taking a unique and potentially incompatible, approach for each one. Our purpose is not to extol technology, or sell you on its benefits, it is to help you understand how to apply it and understand the issues in the use of laboratory automation technology.

This book is important to you because the success of laboratories depends on how well they manage and use the data and information that is collected. That in turn means that use of information management technologies has to be viewed from a strategic perspective—*a perspective that has to be adopted by individual scientists and technicians as well as managers and directors.* Without that viewpoint, possibilities are all that we will get from one of the most rapidly developing technologies. This is not a book of answers, it is a book of questions whose purpose is to lead you to the answers that mean something in your particular situation, and thus will be of value to you. The material between the covers will become yours only insofar as you apply it.

This material is going to look like an exercise in systems engineering, something lab personnel feel is better left to an information systems (IS) group than to scientists and technicians. Lab personnel, in all types and phases of laboratory work, need to be fully aware of, and take responsibility for, the use of computing facilities in their work. The implementation may be done in cooperation with IS professionals, but the responsibility is the labs. Laboratory sciences are as much a matter of data and information as they are about experiments. Understanding how to apply the tools of data handling has become as much a part of the scientists skill set as doing a titration, conducting a clinical study, unraveling a gene sequence in DNA, or conducting an experiment in high-energy physics. Computer literacy once meant knowing how to turn a computer on, use a work processor or spreadsheet, or do data acquisition. Now we need to be cognizant of effects and interplay of data systems, how information moves, is formatted and managed. Lets consider this in light of the development of automation on manufacturing assembly line.

Before any power tools or automation existed, assembly line work was a matter of skill and muscle. Parts moved from one station to another as each section's work was completed. The introduction of power tools (drills and wrenches, for example) would speed up work in stations where those tools were useful, but have no positive effect on other stations. What would happen is that one group of workers, because their efforts were assisted by power equipment, got their work done faster and caused bottlenecks to appear downstream in groups that were still on a manual mode. Real gains in efficiency and effectiveness wouldn't occur until the entire assembly process was looked at as a single system, and automation applied so that the output of one station would move smoothly into the input queue of another. If a station's operations could not be immediately improved, then multiple copies of that station were created to handle the material flow.

That same process goes on in laboratories today. Specific tasks in a lab are automated through the use of intelligent instruments, robotics, and data systems, with little or no coordination. Then people decide that a manual transcription process of moving data from one place to another needs to be replaced with an electronic one. In quality control labs, this usually occurs when a Laboratory Information Management System is being considered. What is discovered at that point is that integrating those pieces of a lab automation puzzle is going to be expensive and difficult, because each instance of automation was considered as an isolated case, and not part of a larger process.

This situation is becoming more of a problem as validation of laboratory systems is attempted on a formal basis (validation has always been part of the scientist's responsibility; the formalization of documentation and regulations are the new pieces). Managers are faced with having to validate a large number of steps because the entire process wasn't thought through from the start. Planning for validation is part of the systems engineering viewpoint. Helping managers, scientists, and technicians understand and use that 'systems engineering' viewpoint is what this work is all about.

Joe Liscouski

Contents

PART 1 DEVELOPING A COMPUTER STRATEGY

1 INTRODUCTION 3
- Current State of Computing Technology / 4
- What Key Elements Have Changed Over the Past 10 Years? / 8
- What is the Effect of These Changes? / 12
- What Do We Hope to Accomplish in This Book? / 16

2 A PROPOSED MODEL FOR LABORATORY WORK 18
- Problems with Earlier Models / 19
- The LASF Model / 21
- References / 35

3 MOVING TOWARD A STRATEGIC VIEW OF COMPUTING TECHNOLOGY 36
- The Need for a Planned Approach to the Use of Computing Technology / 37
- Human Considerations in Technology Planning / 38
- An Approach to Planning and Implementing Computing Technology / 46
- Forming a Corporate Computing Strategy / 47
- Summary / 59

4 DEVELOPING AN IMPLEMENTATION PLAN 62
- The Implementation Plan / 63
- Looking at the Movement of K.I.D. / 64
- Communications between Groups / 77
- Setting Criteria for Evaluating Products and Technologies / 84

- Planning for the Introduction of Revised/Update Software / 86
- Make or Buy Decisions / 88
- Testing the Implementation Plan / 90
- Of Particular Interest / 91
- References / 103

PART 2 TECHNOLOGY FOR LABORATORY AUTOMATION AND COMPUTING

5 TECHNOLOGIES FOR IMPLEMENTING THE LAB AUTOMATION MODEL   107

- Sample Storage Management / 108
- Sample Preparation—Laboratory Robotics / 111
- Instruments and Measuring Devices / 119
- The Data Librarian / 127
- Initial Considerations for Implementing the Librarian / 129
- Laboratory Information Management Systems / 132
- Electronic Laboratory Notebooks / 137

6 COMPUTING AND INFORMATION TECHNOLOGIES FOR THE LABORATORY   142

- Operating Systems / 143
- Data Storage / 150
- Client/Server Systems / 156
- Increasing Computer Power / 159
- Hyper-Information Systems / 164
- Networks and Communications / 169

APPENDIX AIA CHROMATOGRAPHY DATA STANDARD SPECIFICATION   177

INDEX   205

*Laboratory and
Scientific Computing*

PART 1
DEVELOPING A COMPUTER STRATEGY

CHAPTER 1

Introduction

The use of computing technology is going to continue to change the way we work. In laboratory work it has progressed from a tool used by the daring few in the 1970s for instrument control and data acquisition to a commonplace add-on to most instrumentation. In research, mathematical modeling and computer simulation are now just two more tools to work with.

Changes will continue, and we will all continue to be affected by it. Rather than just being concerned about it, we need to plan for it. The level of planning you do is going to be a major determining step in whether information, computing, and robotics technology becomes a valued tool or swamps you. That planning doesn't mean that you have to be an expert in computing. Rather, forget for a moment that computers exist and give some serious thought to how your work is currently being done, and how you would prefer it to be done. Then determine how information systems can help.

The point of this book is to help you in that determination. There will be discussions of computing technology, not at the "bits-and-bytes" level but rather what the technology is and what it means to you. A carpenter may want to cut a pice of wood or make a joint when making a cabinet, and so needs to understand how to plan the project and choose the tools that will be used. Understanding how a tool is made may help appreciate its use and quality, but that knowledge isn't necessary when choosing between a drill or a saw. That's the direction this book will take. The emphasis will be on how to plan, what the tools are, and what they mean to you. It would be helpful (*but not required*) if you sat at least once in front of a system and used it. The concepts discussed in some sections will mean more to you if you do so.

Many of the examples in this and the following chapters will be taken from situations in chemistry. That doesn't mean that this is a book for chemists or just those in the chemical industry. From the standpoint of computing, the issues are the same regardless of whether your discipline is biology, physics, mathematics, architecture, and so on.

Note: In various places mention will be made of particular products. This should not be taken as an endorsement or recommendation, but simply as an example. At the current rate that software and hardware are changing, any comparison or relative statements of capability can become outdated very quickly.

CURRENT STATE OF COMPUTING TECHNOLOGY

The current model for laboratory and scientific computing is one of interconnected machines providing both local resources for graphics display, human interaction, real-time data acquisition, and networked access for storage and high-performance computing (Figure 1-1). Los Alamos National Laboratories *Computing in the 1990's* views the desktop system as a window into computing resources available anywhere in the world.

Today many researchers have access to communications networks that span countries (Figure 1-2 is an illustration of NSFNET as it existed in 1988) and continents. The InterNet, with access through government, educational, research, and commercial sources, is the forerunner of the Information Super-highway and provides users throughout the world with EMAIL and file transfer utilities. In some cases those networks are funded by government agencies. In others they are private networks put in place by companies to support their businesses. The dependence of research on networked communications is being addressed by the government in the form of the National Research and Education Network. In its 1987 statement of the "High-Performance Computing Strategy", the recommendation reads "U.S. Government, industry and universities should coordinate research and development for a research network to provide a distributed computing capability that links the Government, industry, and higher education communities". Figure 1-3 shows the timetable for that work, and Figure 1-4 shows how that capability might be used (from *Grand Challenges: High Performance Computing and Communications*, A Report by the Committee on Physical, Mathematical, and Engineering Science, Supplement to the President's Fiscal Year 1992 Budget).

Banking and insurance companies, stock brokers, the health care industry, and other information-intensive institutions have similar computing and communications models that lag the scientific use by only a few years. That gap is narrowing quickly. The techniques that were devised to solve numerically intensive applications in science are quickly finding appreciation in other markets.

Figure 1-5 (from *The Federal High Performance Computing Program*) shows part of the hardware requirements needed to support the "Grand Challenges" of science. The business community could add its own computing requirements for economic modeling, real-time analysis of point-of-sale data to spot market trends, and other applications.

Computing technology has progressed to the point where the hardware and software are truly commodity items, yet the effective use of this

CURRENT STATE OF COMPUTING TECHNOLOGY

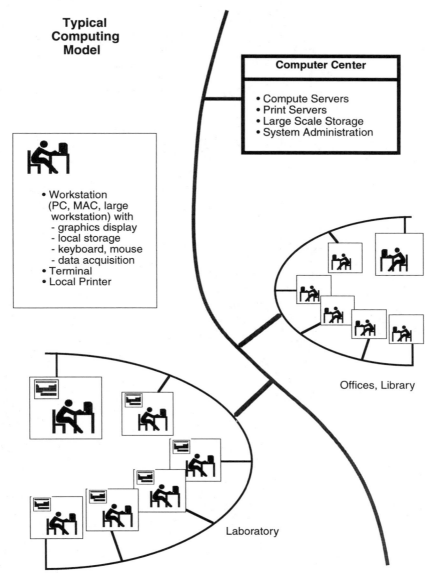

Figure 1-1 Typical computing model.

technology is still in the future. It's not a matter of the sophistication of the tools. It's more an issue of applying them and doing so in a way that overcomes the increasingly rapid development of new products and services.

"Nearly one-third of Americans have a computer at home, work or school, but most use them only as toys or high-tech typewriters..." (*Boston Globe*,

Figure 1-2 Simplified view of NSFNET in 1988 (National Science Foundation Network).

week of March 25th, 1991, by Tim Bovee, Associated Press). That problem isn't limited to the home or business office. Those working in the sciences, where computing technology is routinely pushed to its limits, also exhibit some of the same characteristics. A visit to one company found the following situation:

> The organization had recently spent a large sum of money on computers and networking to upgrade the tools available to its researchers. During my visit there were three types of applications in frequent use: word processing, data acquisition, and games. A report would be prepared by a researcher (edited, corrected, revised) using a standard work processor package and then printed in its "final form". The report would then go into the company's library system.

CURRENT STATE OF COMPUTING TECHNOLOGY

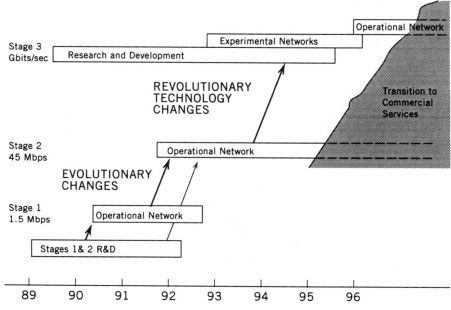

Figure 1-3 Timetable for the National and Education Network.

Because the system used by the researcher was not compatible with the system the library used (both on the same network!) the report had to be *retyped* into the library system. The data acquisition systems were built in-house. One was inoperative because the designer had left the company and no one knew how to use it. Another in-house development was based on eight-year old technology that was only understood by one individual.

Other companies share similar problems. Departments often duplicate work because they don't know that the data already existed or cannot access it. Companies need to recognize that their intellectual wealth exists as knowledge, information, and data, and that laboratories are prime contributors to that pool of wealth. The development and implementation of corporate computing and information strategies that actively incorporate the products of research and support laboratories put a company in a better position to gain full benefit from its investment in R & D. Those that do not, will find themselves at a serious competitive disadvantage in an economy that depends upon a fast response to change, flexibility, and the ability to capitalize fully on every dollar of expenditure.

The current emphasis on cost-cutting to improve profitability will run it course. Long-term, sustainable profitability will depend on the creation of new products. Support for those product development efforts will be based on a foundation of laboratory data, information, and knowledge. If that foundation

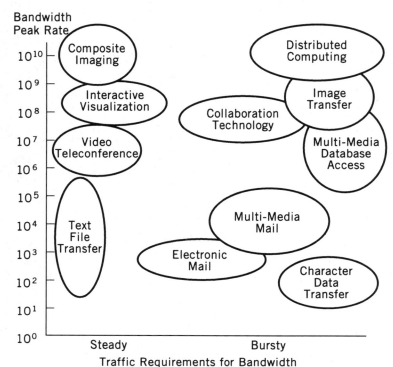

Figure 1-4 NREN applications by bandwidth and traffic characteristics.

is weak or nonexistent, R & D will incur the cost of duplicate expenditures to redevelop data that couldn't be found. Those expenditures in time and money may show up in reduced patent lifetimes, excessive development costs, and possibly legal costs in defending patents and supporting product claims.

The potential for the use of computing, information and communications technologies is far ahead of its actual use. These problems are not the result of poorly trained people, but rather the lack of a coordinated approach to using technology.

WHAT KEY ELEMENTS HAVE CHANGED OVER THE PAST 10 YEARS?

About 1980 an event occurred that changed our perceptions and expectations of computing. That event was the development of Visicalc, the forerunner of today's spreadsheet programs.

Prior to its development the use of computing technology was dependent on programmers to develop applications to meet the needs a specific end-user or a narrowly defined segment of the overall marketplace. Most software was

WHAT KEY ELEMENTS HAVE CHANGED OVER THE PAST 10 YEARS?

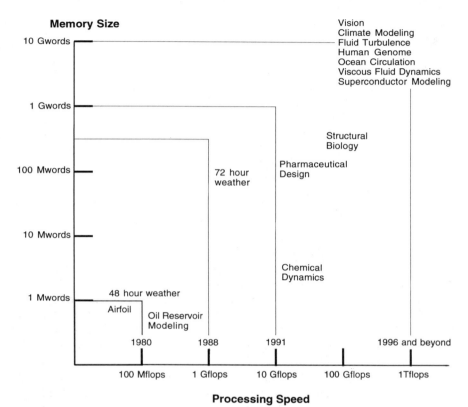

Figure 1-5 Hardware requirements.

written with the expectation that the users had some knowledge of computer systems. They may not have been programmers, but they had the ability to get the system started, which required some understanding of how things worked. Visicalc demonstrated that there was a mass market for computer software that didn't depend on having ready access to people who knew Fortran, Basic, Cobol, etc. For those of you who are sitting and saying "it was the microchip....", I would counter that the economics of the microchip helped realize the potential that Visicalc offered. Without the need created by mass market software, the chips would have much less effect. The results of mass market software are evident today in any shopping mall; you can shop for computers and software the same way you shop for stereos, audio tapes, and, audio compact disks.

Dr. Abraham Maslow developed "Maslow's Pyramid of Human Needs" (Figure 1-6). That diagram prioritizes basic human drives. A higher-level need cannot be managed until those below it have been satisfied.

The same structure can be drawn for computer systems with "hardware capability", "software capability", and "integration" as the driving factors. That

Figure 1-6 Maslow's pyramid of human needs.

diagram (Figure 1-7) not only prioritizes system requirements, but also gives us a broad-brush view of stages that computer technology has gone through.

The period up to 1984 was primarily concerned with hardware capability: raw computing power, high-performance graphics displays, and physically smaller systems. The larger computer vendors began providing scientific and engineering workstations (the dominant feature was the large, high resolution graphics display). IBM began the personal computer market with microprocessors, a simple operating system, and minimal graphics capability. There were a number of other vendors providing low-cost systems, but IBM's entry "legitimized" the market. The availability of those low-cost systems did two things from the vendors' point of view in the mid-1980s: it removed pricing as a major issue for people purchasing computer systems, which opened a

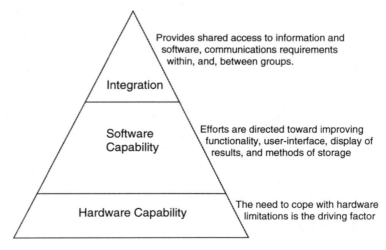

Figure 1-7 Pyramid of computing needs. (© LASF)

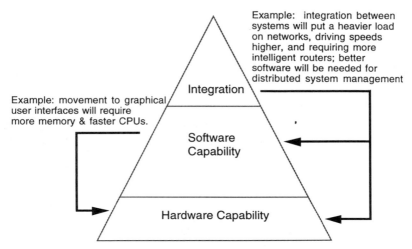

Figure 1-8 New developments in software and requirements for integration are going to drive new technologies at the lower levels. (© LASF)

potentially large market, and it gave software developers a standard platform on which to build applications. That shifted the market from the hobbyist to the general public. From the end-user's perspective, software (the tool needed to do a job) became affordable and choices were available (a competitive environment was created that would limit cost and improve capability)—we began to move to the second level.

Just as standardized hardware packages and operating systems (MS-DOS and the Macintosh system) made mass market software possible, standardized operating systems are going to make integration possible—but we're not there yet. There are at least six common operating system environments in use today: MS-DOS, the Macintosh system, MVS on large IBM systems, VMS on Digital Equipment's VAX, and, UNIX (UNIX is a trademark of AT&T)— Microsoft's Windows NT. IBM's OS/2, while technically a sound system, has yet to achieve wide acceptance. There are alternatives from Hewlett-Packard and others, but those are the dominant ones. Each is a "standard" in that it has wide use and acceptance, and the installed base of equipment is large enough to support third-party software and hardware companies. UNIX (in all its flavors and versions) and MS-DOS are "industry standards" since hardware systems can be either built or purchased from a number of vendors and run the same software. In both cases this "standardization" has been the result of user acceptance rather than industry legislation.

Standardization of the operating system has a number of implications. People can migrate from one computer to another (from another vendor perhaps) and work without retraining. The file structures—the way information is stored—will be compatible so that applications software and information can move from system to system. And finally, network software can be built on a common base, making communications between systems simpler.

We are now (1994) in the beginnings of the third level—integration. There are some excellent examples of application and systems integration today—depending on what you mean by "integration". Among the best examples for an integrated user-interface is the Macintosh (Apple Computer Corp.), which has a consistent behavior across applications from Apple and third-party developers. Oracle (from Oracle Corporation) is a good example of database integration; applications and databases can move between computing platforms with little or no change.

If "integration" means the ability to move software and information between computer systems, without regard to vendor, and have them work, we're not there yet. Individual vendors are approaching the problem of integration in their own way, usually with *their* systems playing the central role in making everything work. Digital Equipment, Apple Computer, and Microsoft (through Windows extensions) are making strong efforts to accommodate other vendors.

The movement through these layers is not a one-time process. New developments in hardware capability inspire new software. User requirements, usually expressed and satisfied through software, put more demands on the hardware for more speed and capability at lower prices. Integration efforts will drive both hardware and software developers to be more innovative and provide standardization, if for no other reason than to reduce the cost of system development, maintenance, and pressures from third-party vendors for compatibility and transportability.

One of the by-products of the market's competitive pressures on software development has been the rapid, almost panic, development of operating system versions. A third-party vendor provides a useful add-on to a popular operating system, and that OS's next version incorporates it or a look-alike. Things are moving fast enough that the consumer becomes the test bed for large-scale debugging of systems. The OS vendors seem to have forgotten the difference between an operating system and layered software.

WHAT IS THE EFFECT OF THESE CHANGES?

The primary result of these changes is our willingness to view computer systems as tools and use them. Computer systems, networks, and database systems no longer invoke the fear of breaking new ground as they did 10 or 15 years ago—particularly when a twelve-year-old can flip a switch to get access to more computing power than an engineering department had in 1980!

WHAT IS THE EFFECT OF THESE CHANGES?

Over the past decade computing systems have become faster, less expensive, and easier to work with. Enough software has become available that most people using computers will never think about programming new applications, at least not in traditional terms; graphical user-interfaces on database programs, for example, allow users to create new applications by pointing-and-clicking, almost "drawing the application" by visually creating relationships. One example comes from laboratory data acquisition. In 1978 Digital Equipment Corp. produced the MINC (Modular Instrument Computer), which was based on the PDP 11/03 and later the PDP 11/23. A brief comparison of the hardware specifications between it and a Macintosh running National Instruments LabVIEW 2 is given in Table 1-1.

Comparing technologies that are 13 years apart isn't particularly fair to the MINC (which in its day was a good value), but it does show how things have progressed. Almost every job that the MINC was put to use for had to be programmed in Fortran, BASIC, or Macro-assembler. While people did very clever things with the graphics, the human interface was very limited and consisted of a lot of typing. Though the hardware specifications of the Macintosh configuration are awesome by 1978 standards, they are run of the mill today. What you get for that almost hundredfold increase in memory is improved graphics output, a graphical programming environment (no code, just connect the dots and blocks), and faster implementation and adaptation from people less knowledgeable about computer languages. In short, more capability at lower dollar cost, but a lot more hardware needed.

To get an idea of how things have progressed, Figure 1-9 shows a diagram from Strawberry Tree Incorporated's Workbench MAC software for data acquisition. This simple diagram takes a single analog input and displays it on a meter and as a strip chart. The design and implementation took about 3 seconds. Similar project can be designed with other tools such as Labtech's Notebook for Windows. The same project on mid-1970s vintage machines would take hundreds of lines of data acquisition and graphics coding, plus a lot of debugging. MINC Basic would have done the acquisition and chart display with a single line of BASIC programming, but that was a unique situation (anything more complex would have become very large very quickly). The data trace is from a random number generator used as a simulated input.

The next examples show that the programming complexity for traditional

TABLE 1-1 Hardware Specifications

	MINC (1978)	MAC + LabVIEW2
Main memory	64 kilobytes (or characters)	5120 kilobytes
Mass storage	<10 megabytes	>40 megabytes
Graphics	Very limited, single waveform	Full-color, full graphics display
Approximate cost	~$20,000	~$10,000

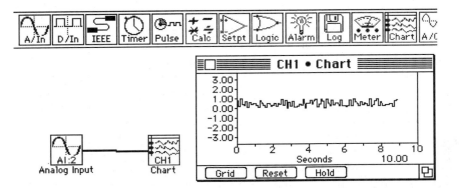

Figure 1-9 Meter and strip chart display from Strawberry Tree Incorporated's Workbench MAC software.

procedural languages can increase significantly. Buffer management and task prioritization can become issues for the software developer. With the graphical interfaces provided by Strawberry Tree, National Instruments, and Labtech, the effort is trivial and adds only a few seconds (seconds!) to the effort. The first example (Figure 1-10) adds file output to problem, and the user has control of data formats.

Figure 1-11 shows the calculation and display of the last three seconds of data as the experiment runs.

Prior to today's products, particularly in the sciences, the major unmet need was raw computing capability. This was most notable in data acquisition, where the machine's speed was a limiting factor in people's work. That problem has been solved in two ways: machines have become faster so that they are better able to handle demanding tasks, and they have become cheaper, which means that large problems could be distributed over several less expensive computers rather than putting too much on a larger system.

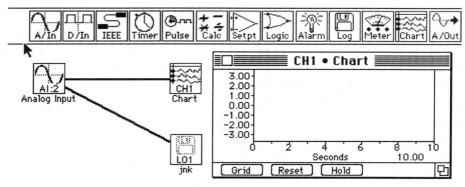

Figure 1-10 Addition of file output to display of Figure 1.9.

WHAT IS THE EFFECT OF THESE CHANGES?

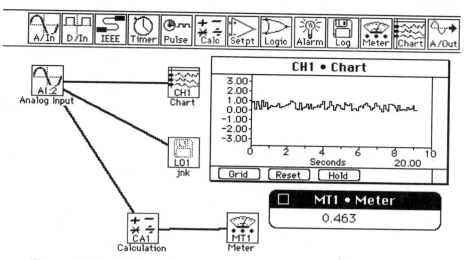

Figure 1-11 Addition of calculation and data to display of Figure 1.10.

Instead of having 16 instruments tied to a system, you use 16 systems (which drastically reduces the cost of software development and management). If you spend all your effort trying to deal with hardware issues, you won't get to the next level. This is still true today in some applications such as high-energy physics, which is heavily burdened with data acquisition and analysis problems, and the realm of simulation and modeling, where raw number crunching power is the most important factor.

Once your hardware constraints have been removed, the next need is software. The control of a program used to be directed by typed commands, now point-and-click interfaces make the software easier to understand and use. Not only has the user-interface improved but the options for output (visualization systems) make the results more useful and easier to understand. Having removed hardware limitations and given the software to get the analysis work done, you can move your attention to the next issue: integration. This is using the output of one system as input to another, making information more useful and usable.

More people, many of them not very knowledgeable about the ins and outs of computer systems, are going to have a lot of capability at their fingertips (or mouse clicks); capability to do more and better work and spend their time being more creative and productive. That has been the promise of every appliance since people stopped washing clothes on rocks. In the case of computer systems it can be quickly realized, or just as quickly frustrated as a result of poor planning.

The availability of power tools such as a radial-arm saw from places like Sears made it much easier for part-time carpenters to make good square cuts and then build furniture or whatever they needed. For others it was quicker way to turn good lumber into kindling. Consider the potential that properly planned and used single-user and networked computer systems can have in an organization and the potential for disaster that can occur if they aren't well planned.

While most of this seems to be going on around you, you may feel you have little control over what a vendor does or how standards are developed. In fact, you have a great deal of control and influence, some of it is direct and some is through software developers.

Through planning and being aware of the options that are open to you, the organization and introduction of computing technology can be a rationalized process. This isn't an activity that assumes you are starting from scratch, very few people are in that position, most have a considerable investment in computing. Most of the pieces you have in place can be accounted for in the implementation phase of the process. It's not quite as easy as rearranging Lego blocks, but computer systems built on widely used hardware and operating systems are flexible.

Having that plan puts your vendors (including internal service groups) on notice that products and services are going to have to meet some defined requirements, and those fundamental requirements are organization-wide (it should also allow room for experimentation and innovation). Vendors will usually respond well to that approach since it gives them a defined working environment, a framework that they can test product capabilities against. It also gives you a basis for evaluating hardware and software.

WHAT DO WE HOPE TO ACCOMPLISH IN THIS BOOK?

There are three points that this book will cover:

- how to go about *planning* for the use of computing technology,
- how to *implement* those plans, and
- how to *assess technology*: what is available today, and what is likely to become important in the near term.

The intent is not to present a prepackaged plan, but rather to provide you with alternatives that need to be considered in formulating your own plans and then putting them into practice. The technology assessment will be a review of computing techniques at a general level and what they could mean to you. The key to working with computing is to allow for as much flexibility as possible and still get something useful done.

In the following chapters we look at the process and issues in developing a computing strategy and implementation plan. There will be some issues raised that do not, today, have resolution. The laboratory notebook is one example. The resolution of that problem lies as much in the legal and regulatory system

WHAT DO WE HOPE TO ACCOMPLISH IN THIS BOOK?

as it does in computing. Being aware of the issue is a necessary step toward its solution.

In the remaining chapters we will look at the technology available today and that which could become promising in the near future. These points will be addressed by looking at technology from desktop system and then expanding into multi-user and networked environments. I hope that this approach will make the choice and trade-offs clearer, as well as clarify the benefits of the more complex technologies.

CHAPTER

2

A Proposed Model for Laboratory Work

In order to discuss *laboratory automation*, much less a strategy to do it, we need to establish a common framework or model. This model should help us:

- Identify a particular topic for discussion and show how it relates to other work that is in progress.
- Develop an understanding of how laboratories function and how one type may differ from another, as well as highlighting their similarities.
- Design the implementation of laboratory automation systems without limiting or predetermining that implementation.
- Position and compare products.
- Understand the need for and placement of standards.
- Highlight where appropriate technologies for automation exist, where technologies are missing, and where conflicts exist.

Above all else, the model should be useful and accurate.

This chapter is a description of one such model. It is the result of a fair amount of effort over the past 18 months and has some of its roots in earlier publications [1,2,3]. Models and descriptions of systems are key means of communicating concepts. There are a large number of people working in laboratory automation who have never worked in a lab.: information systems professionals, consultants, systems integrators, and others are active in the field. This model should provide a useful basis of discussion and clarify system concepts for those not involved in the day-to-day operations of a research or testing laboratory.

One aspect of this work is an emphasis on the process aspects of laboratory work, and the point that the process of carrying out laboratory work is part of a larger process: the functioning of company, manufacturing site, academic institution, or research organization. The justification for laboratory automation projects is coming under tighter scrutiny from management. It needs to include statements of how a particular project relates to other work and how it contributes to meeting lab and corporate goals. This was part of the message delivered by Dr. Robert McDowall at the 6th International LIMS conference [4]. The linking of the laboratory process to that of its "parent" organization is a key benefit of the model we are describing. The nature of the links and the linkage points will vary from lab to lab. Making these connections will help others understand the role of lab work and the value of automation projects. It will also provide an opportunity for making those connections more effective. The electronic transfer of results between groups, rather than printed reports, may make the results easier to use.

The electronic linking of quality control (QC) to process control and inventory management has the potential for improving the production process and speeding up the turn-around of material through inventory. Suppose as a result of improving the reporting of lab data and the linkage between QC and inventory control, you could reduce the time material spent in inventory by a day. Material is shipped faster, the size of inventory is reduced, and the facilities needed to house it can be smaller. Plant construction costs on a new or refurbished facility can be reduced as would the overall operating expenditures. Tying the laboratory into the organizations infrastructure and improving its effectiveness can help gain support for laboratory automation projects.

Some of you may have seen the model before. It has been used at PITTCON 92, Zymark's 1992 International Symposium on Laboratory Automation & Robotics, the 1992 ALEX Conference, and an LASF short course titled *A Strategic Approach to Laboratory Automation* at the 1992 Scientific Computing & Automation Conference and PITTCON 94. The model is consistent with the approach the Laboratory Automation Standards Foundation (LASF) has taken to the development of an information strategy within the laboratory and to the development of links to research and process control, administration, and other parts of an organization. A complete electronic implementation of the model isn't possible today—that will be covered in more detail later. Portions of it are a reflection of how the laboratory could operate, rather than being limited by existing situations.

PROBLEMS WITH EARLIER MODELS

The typical lab automation model is shown in Figure 2-1. This diagram has been used in various forms for over 10 years to illustrate the computing environment. It is the type of model referred to in Chapter I as the typical laboratory computing model. There are a number of problems with it:

- It is a hardware view of a computing environment, showing where tasks take place, rather than the process of laboratory work.

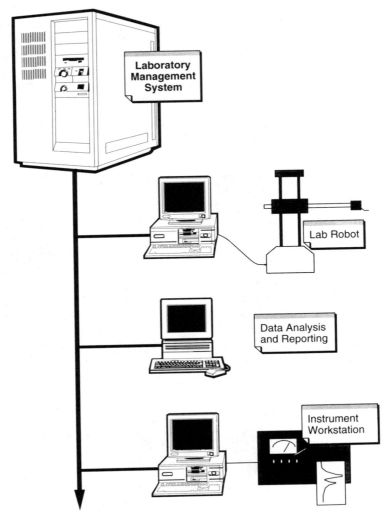

Figure 2-1 Typical lab automation model. (© LASF)

- It is an implementation diagram rather than a model: it assumes a networked structure, for example, rather than letting you decide the best implementation for your lab.
- It shows a network path for the movement of files, messages, and data packets, rather than the exchange of data and information between applications. The ability to copy a file from one computer to another does not mean that the file can be used.
- It doesn't allow you to show the relationship between applications, the need for, or the use of, data exchange standards, nor does it take into

THE LASF MODEL

account the events that initiate and control the movement of materials, data, and information. It doesn't contribute the development of an integrated system, since it fails to show the critical elements that have to be integrated.
- Finally, it doesn't deal with the process characteristics of lab work or help you show the relationship between cooperating groups within an organization.

Models like the one in Figure 2-1 stem from late 1970s and early 1980s work in laboratory automation. Networks were slower, more expensive, and less reliable. As a result they served as a transport for reports and data that was not time-critical. One station would be responsible for a complete process of data acquisition, analysis, and reporting. This information would be sent to another system for inclusion in a Laboratory Information Management System (LIMS) or to another larger-scale data analysis project. Functions were compartmentalized at the data station level.

This level of compartmentalization was convenient for vendors who, because of systems design constraints, have to sell complete data-acquisition-to-reporting packages. Their software and hardware resided in a particular box, carried out certain functions, and related to other boxes in a particular way. The connections—both physical and logical—were left as an exercise for the user. Today we are dealing with the residuals of 1980s design constraints, a factor that is ready for change.

THE LASF MODEL

The model described in this section is shown in Figure 2-2. It is a process view of laboratory automation, showing the steps in the process and their sequence. The controlling factors are added in Figure 2-3, but that's jumping ahead too far. The model can be implemented without the use of computers, all on one system, or anywhere in between—it emphasizes behavior and function, not how to build it.

Four Major Segments

1. *Material Handling and Management.* The primary activities are the management of samples (storage, retrieval, maintaining chain-of-custody, and so on) and their preparation for testing (manual methods, robotics, and others). The term *sample* can refer to manufactured materials, environment specimens, test subjects (animals, plants), and so forth. The components of this segment are
   - Sample Storage Management. This provides physical storage for samples, a means of finding and retrieving them, and laboratory-

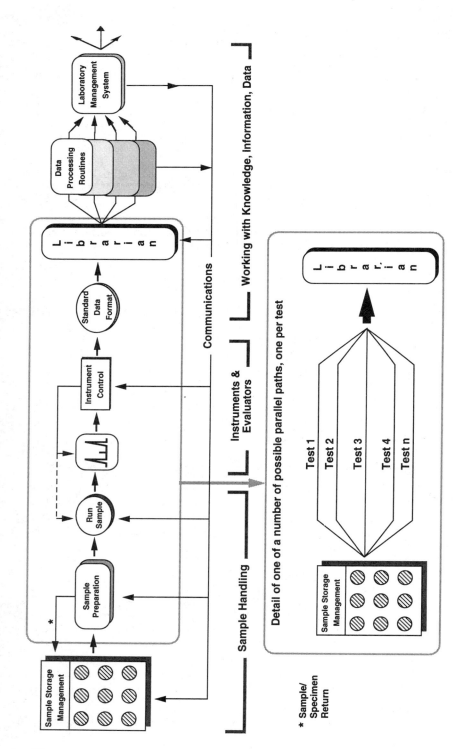

Figure 2-2 The LASF model. Used with permission of the LASF. (© LASF)

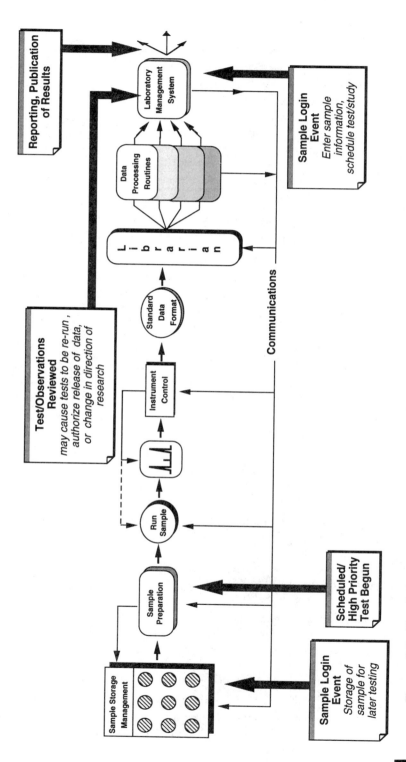

Figure 2-3 The LASF laboratory model with controlling events. Used with permission of the LASF. (© LASF)

specific coding schemes (identification numbers for example). This can be a manual system or an automated facility.
- Sample Preparation. This is whatever activities are required to make the sample suitable for testing.
- Running the Sample. This may be part of the same equipment used to prepare the sample or it may be an independent means (a person, auto-injector, and so on) of getting the sample into the test queue.

2. *Management and Use of Knowledge, Information, and Data.* Here we are concerned with the description of the sample in terms of measured properties and composition, as well as observations—any qualitative or quantitative term. We are no longer working with the physical "sample", but with its description. In addition, this section includes descriptions of test methods and procedures, study protocols, models, relationships, and the laboratories "knowledge base". The latter can be documents (electronic or otherwise) or people's skills or both. The components of this segment are

- The data system used to record and analyze the test results. This can be a computer or a person. It also includes the initial database used to store recordings (electronic or otherwise).
- A standardized format for the movement of observations from the initial database to a library designed to manage all observations.
- The librarian, who is responsible for storing, managing, archiving, retrieving, providing copies of observations, and so forth.
- Data processing libraries, which are collections of procedures used to process observations taken from the librarian. These may be duplicates of those in the primary data system, or completely independent routines.
- The laboratory management system, which can be a person or persons, or a software system, responsible for collecting and managing information, testing schedules, sample status, and so on. It may include an inventory function and provide links to other organizations. In automated QC laboratories, for example, this is the LIMS. In a research lab, this may be the head of the project.

3. *Instruments and Evaluators.* This segment observes, evaluates, or measures the sample. It's the mechanism through which we shift from working with the physical sample to working with its description. Portions of this mechanism may include sample preparation and equipment to introduce the prepared material into the test device—the implementation depends on the requirements of the laboratory.

Note. The sequence of events—sample preparation running the sample, instrument, and so forth—is one of several parallel paths. Each path is

THE LASF MODEL

the process for a particular test sequence. Each begins at the sample storage, and each terminates with the creation of a "formatted packet of observation" being sent to the librarian.

4. *Communications.* This is the mechanism through which the preceding segments and components are coordinated. It may be electronic, a healthy set of lungs, or a combination of the two.

If the language in this list seems odd or as though I were trying to avoid something, I was. The point was to develop a model of behavior that did not immediately assume a computerized system. It is also important to devise a model that is not limited to a particular type of laboratory, but describes work in both research and manufacturing as well as a variety of disciplines. If that has been done and your goal is to create an automated laboratory, then the question is how to implement the model using computer systems.

Once we begin viewing the work in the laboratory as a process, rather than a collection of discrete events, we will be in a better position to improve its effectiveness. The process can be evaluated on the basis of quality of results, turn-around time, and cost-effectiveness. This should not be taken as an effort to diminish the science being done. It is rather an effort to understand how the science is being performed. This viewpoint does have an immediate benefit: once accepted it requires us to keep the process characteristics in mind as we implement each step. At each point we need to consider not only that element, but its relationship to those preceding and those following. In particular we need to understand the purpose, benefits, and mechanism of communications between steps. These considerations are fundamental to creating integrated systems and avoiding the development of isolated automation systems—the "islands of automation".

The process view of lab work gives us the ability to apply process analysis techniques from other disciplines. Whether you're working in research or routine testing, there is a "production" mode to laboratory work: a consistent approach and methodology of developing data from specimens/samples under test or evaluation. Process modeling techniques, which have worked well in manufacturing applications, can be applied to these portions of laboratory work.

Events Controlling the Process

Processes do not run just because they exist. In order for a process to function, and control points are needed to make sure that the process starts and is working properly. Figure 2-3 shows the following events added to the model:

- *Sample login.* This causes activity in the sample storage management system and laboratory management system. It also causes tests to be scheduled based on priority or the needs of a particular study.

- *Scheduled and high priority samples tests are begun.* This stimulates the sample storage management, sample preparation, and movement of prepared samples to test sites (instruments). Tests are performed, and observations recorded. Post-test processing, library storage, and entry into the lab information system are possible subsequent events, based on an examination of observations. This examination may cause retesting.
- *Test observations need to be reviewed.* This can cause samples to be retested. It can also cause the sample storage-management–preparation–running tests-sequence to begin at the start, or any place in-between depending on the reason for retest and the lifetime of the prepared sample. Approval of the results will cause updating of the management system. Partial or completed results may initiate a reporting event.
- *Results/status information is reported outside the laboratory.* This is a critical step. The process operating inside the laboratory is not divorced from processes operating outside of it. Quality control laboratories exist to support manufacturing operations. Research labs exist to further the economic or academic objectives of the organization of which they are a part. "Independent" research labs are dependent upon their sponsors. We need to ensure that the laboratory process has effective links to those other groups. This is a key and often missed opportunity to gain support for laboratory automation work.

We have a process and a description of the events that control it. How do we know it works? That is the purpose of a "supervisory" process of systems validation. Regardless of how a model is implemented, that implementation should be validated. This is one of the key elements of regulations and recommendations from the Food and Drug Administration, Environmental Protection Agency, and International Standards Organization. The "process" view of laboratory work means that the same methodology used in validating any other process can be applied to laboratory automation.

Potential Developments Based on the Model

The purpose of this model is not just to demonstrate how a laboratory operates. It can be used to illustrate the need for data exchange standards (each arrow is a place for standards to be developed and applied). It should also help us define where work needs to be done and highlight areas where the available technology needs to be improved or where technology needs to be developed.

The implementation of the entire model depends on the development of standards for the exchange of knowledge, information, and data. The primary concern is with information and data. Standards that encapsulate data in a

THE LASF MODEL

usable format, such as those developed by the Analytical Instrument Association (AIA) for the exchange of chromatographic data, are an excellent example.

The ability to create these standardized data format modules opens several opportunities. The first is the development of a Data Librarian. Work is underway by the AIA and other groups to extend their initial concepts to include other analytical techniques; data formats for mass spectroscopy are due in the Fall of 1994, and work is beginning on optical and atomic spectroscopy formats. Even if the standard was restricted to one technique, a laboratory would have the ability to create hundreds of data modules, each containing the data for a single set of measurements, in a short period of time. How do you manage them?

DATA LIBRARIAN This management is the role of the Data Librarian (note: the Data Librarian doesn't exist as yet, this is one possible description of how it might work: as a result this could come under the heading of science fiction, the first I'm admitting to).

The typical hardware implementation of a laboratory automation system consists of several data stations, each connected to one or more instruments (Figure 2-4). Since each data station has its own disks and file structures, it is very likely that data files with the same name will appear on different computers. PC DOS implementations have a limit of an eight-character filenames and a three-character extension. You can calculate a large number of filenames from information (at least 3×10^{14} if you limit the characters to letters and numbers), but most filenames are generated on the basis of a logical construction such as a combination of date, sequence number, and file type, which limits the range and almost guarantees duplication between machines. Just keeping track of files on a single computer dictates the need for a library management system.

The Data Librarian is a database system where all data that is supported by standardized data formats, can be stored, retrieved, archived, and, support historical and ad hoc searches. Each data packet from an experiment would still remain an individual file, but the naming and management of the files are the responsibility of the database software, behaving as a large public lending-library. The major difference is that people would request a copy of data file, not the original. It should also have the following characteristics:

- *It should be scalable.* Implementations (different versions possibly) should run effectively on a typical desktop data stations and in client/server environments. In large-scale systems it should automatically support the migration of data between disk, tape, and optical media, and support CD-ROMs for long-term archiving.
- *Consistency.* The user-interface and functionality should be consistent from one vendor implementation to another. A research project is

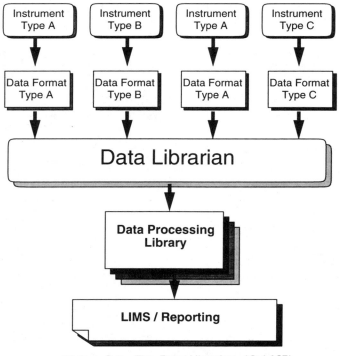

Figure 2-4 The Data Librarian. (© LASF)

needed to develop a set of functional specifications and determine the relevant standards needed to support the librarian.

The use of a data librarian would assist meeting the requirements of regulatory agencies and the validation process.

POST-RUN DATA ANALYSIS Developers and researchers who are interested in new techniques for analyzing data—new peak-picking algorithms or interactive graphics systems for example—now, through data format standards, have a means of accessing data without having to build a data acquisition system. As a result libraries of processing routines can be developed and applied independent of the system used to acquire the data. This provides existing vendors with a new potential market for existing work and room for new developments for analysis algorithms and processing software.

This also offers instrument manufacturers some additional latitude in data station design. An instrument control system need only manage the instrument, acquire the data, and export it in a standardized format. Data analysis, reporting, interactive graphics functions, and other post-run facilities become part of a separate application or group of applications. These applications may be provided by the instrument manufacturer through the modularization of larger existing packages or through alliances with third-party developers. The benefits are increased flexibility in data system design, a potential reduction in support and development costs, and the ability to choose how deeply a vendor wants to dive into the development and support of computer systems. The users of these systems will benefit from better designs for integration, more flexibility in implementation, and, potentially, a wider range of choices.

This separation of functions, particularly the removal of data formatting from the analysis procedure, also provides another benefit. With standardized data formats, and standardized links into a data analysis procedure, we can develop and use libraries of real and synthesized data files. Those libraries can be used to test and compare different analysis algorithms.

This will have an impact on the standardization of test procedures. The test method can specify sample preparation, data acquisition parameters, data analysis algorithms, and reporting formats without specifying a particular vendor. The specification can include baseline correction procedures, methods for development of calibration curves, and data reduction routines that work best for a particular problem. Each portion is a separate set of algorithms that can be linked together, similar to UNIX pipelines, to create an analysis package. It should be a simple matter to adapt the graphical interfaces used by packages like LABview, Labtech Notebook, and others create and document complete analysis procedures, from data preparation to reporting. This will considerably improve the development, documentation, and validation of standard operating procedures.

MODULARITY The model (Figure 2-5) stresses modularity. The modularity is useful only if standards are specified to interconnect the components. Each arrow in Figure 2-5 is an opportunity for standards development, and that work should be strongly encouraged. Using the arrows as a guide will clarify the relationship between development efforts, and areas where work needs to be done easier to identify. It leads to major benefits in two areas:

- *Integrated Systems.* Standards and modularity taken together form the bases of integrated systems. Once they exist, the every-system-is-unique-and-a-custom-design-effort school of systems design is history. This is a real benefit to integrators as well as lab managers and scientists. The problem becomes easier to define, the costs easier to manage, and the likelihood of successful implementation increases. This should simplify overall system management and documentation.
- *Validation.* The requirements for validation won't change. The means of satisfying those requirements will become easier to define and implement. Each module can be validated, as are the links between modules.

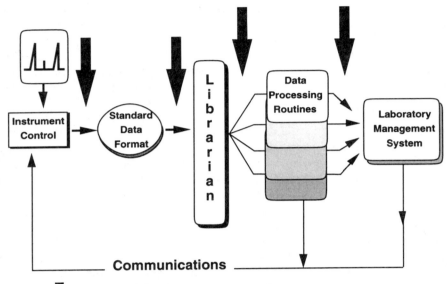

Figure 2-5 Modularity. (© LASF)

If those linkages are based on standards with tested implementations, the effort involved for each individual laboratory is reduced. The modularity provides test points so that each element can be independently examined. As the system changes—modules can be replaced or added—the revalidation becomes confined to the section modified. In principle this is the situation today, it isn't realizable because the modularity and standards don't exist. Large program structures, which are common in custom implementations, require that the entire structure be tested since the effects of a programming change could have detrimental side-effects that may not have been anticipated.

MODULARITY'S IMPACT ON DATA ACQUISITION SYSTEMS Extending the ideas of modularity into the internal workings of data acquisition systems offers a solution to a developing problem. The development of operating systems (OS) for desktop systems increasingly favors the commercial office environment. Where early OSs stressed compactness, speed, and the ability to respond to real-time events, modern systems stress ease-of-use facilities for an office function where the market share is larger. Given the marketing push

THE LASF MODEL

from the major software vendors, this shift to non–time-critical applications is going to continue.

These same developments, plus the growing list of operating system modifiers (specialized screen savers, alarms, on-the-fly file compression, network access, etc.) significantly compromise their use in real-time data acquisition work. One solution is to dedicate a system to data acquisition functions alone, isolating it from anything that could potentially interfere with its primary task, including stripping down operating systems, eliminating screen savers, turning off network access, etc. This is probably a good intermediate solution for research applications, particularly if flexible closed-loop interaction, or interactive acquisition and display, such as that afforded by LABview, Labtech Notebook, and Strawberry Tree's Workbench are important. Another approach, more suitable for commercial data acquisition and analysis products, is to remove real-time data acquisition from the desktop computers operating system environment entirely. This does not necessarily mean that it has to be taken out of box, just logically disconnected from operating system constraints and interaction.

The current structure of data acquisition, processing, and reporting residing in one package on top of a commercial operating system is a carryover from early 1980s implementations as noted earlier in this chapter. Current hardware, consisting of inexpensive processors, large memory buffers, and small, low-cost, low-power consumption disks, makes it possible to develop dedicated boxes (or boards inserted into a bus slot) that carry out all data acquisition functions up to and including formatting the raw data into a standardized data file format and sending the resulting file over a communications link to desktop system for further processing. It should be able to communicate directly with the Data Librarian noted above. The data acquisition conditions and data filename prefix would be sent to the data acquisition system as part of its initialization sequence. The software could be encoded in ROM or down-loaded from a library of routines. These boxes would manage all real-time interaction with the instruments and leave the desktop system available for other work. This approach leaves you free to take advantage of all the ease-of-use facilities in an OS and still get instrument data acquisition and communications accomplished.

Out-board data acquisition systems have been in place in applications before. The Waters LACE box, the Nelson Analytical integrating A/D systems, and the custom interface boxes used in Zymark robotics systems all provided a means of separating real-time interrupt handling from operating systems. The Waters and Nelson systems provided data acquisition, but the results were only used with their proprietary analysis systems. The Zymark boxes handled all real-time interrupts and programming and provided a command structure that allowed the box to be polled by the robot; collected data was held by the interface box until required. That allowed real-time control of a device without interrupting the working of the robot or having to include interrupt and event-handling software in the robot's control algorithms. The principal difference is that the modules described above are complete data acquisition systems that provide their results in a standardized manner for post-run processing on a separate platform.

Since the only released standard is the AIA Chromatography data format, we'll use that as an example. This is how the process would work, excluding sample preparation.

> Once the samples are prepared and inserted into the sample tray, the data acquisition system is initialized. Data collection begins and each chromatogram is stored in a separate file, using a filename prefix plus a counter for the names of the files, and sent to host system for storage through the data librarian. When the analyst wants to generate a report on the day's analysis, he tells the reporting software which sample files to work with, the processing algorithm (taken from an independent library of modules), and the report format and the work is completed. Graphical interaction with one or more chromatograms would be done through a separate set of programs that allowed display, interactive baseline correction and computation. The computed results are sent to a LIMS system.

The difference between this example and current practices is that software and hardware modules can come from one or several vendors. The user can pick and choose depending on what is best suited for a particular situation. One immediately useful result of this methodology is that data becomes immediately available for interlab transfer as soon as it is entered into the librarian. In large installations there may be one data librarian for the entire facility.

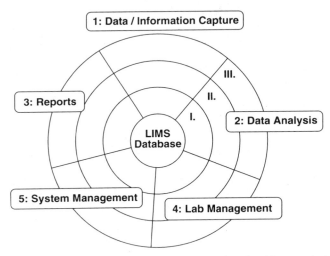

Figure 2-6 LIMS concept model. (© ASTM. Reprinted with permission.)

THE LASF MODEL

TABLE 2-1 LIMS Concept Functions

Level 1 Minimum	Level 2 Intermediate	Level 3 Advanced
Global Issues		
Change control	On-line document	Version control
Documentation	Group security	Test and specification
Quality	On-line training	Revision control
Security	Intermediate interfaces	Security by object
User-interface	Validation tools	Advanced validation tools
Validation	Chain of custody	Multitasking user-interface
	Configuration tools	Multimedia
	Audit trail	Advanced configuration tools
LIMS Database		
Simple database technology	Intermediate database technology	Advanced database technology
	User definable fields	Object based
	User definable indices	Natural language
		SQL
		Client server
		Transaction rules
		Distributed information and processing
1. Data/Info Capture		
Manual sample login	On-line from instruments (one-way)	Bidirectional communication to/from instruments
Manual results entry	File Transfers (one-way)	IR,UV.NMR spectra
	Bar code entry	File Transfers
		2-way links to external systems
		Multimedia/imaging
		Electronic notebook
2. Data Analysis		
Result verification	Comparison of result specifications	Inter/Intra test calculations
Basic calculations	Predefined math functions	Advanced math functions
	Intratest calculations	User defined functions
	Graphical presentation	3-D graphs
	Basic statistics	Advanced statistics
	QA/QC on samples	Dynamic links to prior results and other systems

TABLE 2-1 Continued

Level 1 Minimum	Level 2 Intermediate	Level 3 Advanced
3. *Reporting*		
Predefined Reports	User defined Reports	Natural language reporting methods
Sample labels	Queries, sports, filters	Batch reports
	Basic graphics	Event triggers
		Export to external systems
		Bulk data transfers
		Advanced graphics
		Multisite LIMS reports
4. *Lab Management*		
Sample/order status	Scheduling of lab work	Resource management
Sample/order tracking	Location of sample	External system scheduling work
Backlog report	Workload prediction	Stability (time) schedule
	Pricing/invoicing	AI decision making tools
		Revenue/cost tracking
		Advanced QC management
		Multisite LIMS management
5. *System Management*		
Backup and recovery	Archiving	Dynamic performance tuning
	Manual performance tuning	Advanced systems fault tolerance
	System fault tolerance	Redundant systems
		Advanced communications links to external systems.

This same process can be played out for other analytical techniques using the same librarian for data storage. The only requirement is that the standards needed to contain the data exist and be used. This isn't a scenario that is far off in the future. Elements of it exist today, missing points are a matter of repackaging existing software.

The description above, while focused on chemistry, is appropriate in any scientific discipline. Substitute EKG for chromatograph, and the same behavior applies. The key is standardized data formats.

In this model realizable? Yes! The AIA is expanding the number of standards, so the key element either exists or is under consideration and development. The data analysis and reporting modules exist—they are part of existing data systems—so they need to repackaged. The only major piece that does not exist is the Data Librarian, and that is not a major engineering effort.

Figure 2-6 is taken from the *ASTM LIMS Guide*, the result of work by ASTM committee E31.40. The purpose of the model is to provide a framework for understanding and relating LIMS functions. Level 1 depicts mandatory functions, Level 2 shows intermediate functions, and the third level is for advanced functions and technology. Each level applies to five major segments. Table 2-1 gives a list of each function by level.

The LIMS Guide also includes a detailed description of functions, terminology and data flow models.

A comparison between the ASTM work and that described earlier is a matter of perspective. The two agree, one focuses on the behavioral aspects of the system, the other emphasises the details that need to be included in an implementation. While covering details, the ASTM model does not dictate implementation. Some functions that it describes are appropriate to the Data Librarian, others to a reporting system. In one way it is like comparing perspectives on New York City. A standard map shows the relationship between objects and their interconnections. The ASTM model then is similar to a modified perspective that emphasises a particular landmark, the Statue of Liberty, and relates everything else in the city (transportation, lodging, etc.) from the standpoint of someone whose only central focus was Ellis Island.

This does resurrect the discussion of what a LIMS really is. That discussion will be left for a later section., The problem now is how to use the model described early in this chapter and develop the strategies to make lab automation an effective and efficient process. That model should be altered to fit the needs of your facility. Having a description of how you want the facility to operate is fundamental to determining a strategy to realize that behavior. That is where we move to next.

REFERENCES

1 "Planning an Approach to Laboratory Automation", *Computers in the Laboratory*, ACS Symposium Series #265, American Chemical Society, Washington D.C., 1984

2 Liscouski, J., "Issues and Directions in Laboratory Automation", *Analytical Chemistry*, Vol. 60, #2, January 15, 1988 page 95A–99A

3 Liscouski, J., "Inter-Connectivity", *Laboratory Robotics & Automation*, Vol. 3, pp. 145–149

4 McDowell, R., "A Strategic Approach to Laboratory Automation", presented at the 6th International LIMS Conference, Pittsburgh PA, June 1992

CHAPTER

3

Moving Toward a Strategic View of Computing Technology

Abstract jigsaw puzzles based on the tiling of simple geometric shapes have something in common with the way automation projects are approached. The usual approaches to sorting pieces, starting at the edges and so on, were frustrated because the design of the puzzle was so uniform that once you had a few pieces put together, there was no guide to how they fit into the overall pattern.

Many approaches to laboratory automation suffer from the same problem. Attention is paid to parts of the labs needs, but until an overall plan is put in place, the work done one one project may have no bearing on another piece. Just as the puzzle lacked reference points, computerization projects lacked the necessary overall framework to show how they relate to each other.

The point of this chapter is to lay the groundwork for determining your strategic view of laboratory computing. Rather than try to present a fixed plan, we will look at the guideline for determining one that can be tailored to your facilities requirements.

This planning process has taken on a new urgency as more companies feel the weight of regulatory efforts, particularly ISO 9000. As compliance to standards and regulations becomes an important factor in a company's ability to do business, having a considered and viable process for working with and managing information is going to be paramount.

In most organizations, computing technology has been introduced on an "as you need/justify it" basis. In many cases the justification is hindered by the experience of others using computers; systems may have been purchased and not used effectively, making a request for a new system less likely to be supported. A study released in 1989 by the *National Training & Computer Project* (Raquette Lake, NY) states that "as many as 40% of employees are not using the software that has been purchased for them by their corporations". Systems either weren't being used, or they were used to only a small fraction

PLANNED APPROACH TO THE USE OF COMPUTING TECHNOLOGY

of their potential. Inadequate training, poor software selection, and a lack of understanding of what was needed to get a particular job done contributed to this problem.

Questions concerning the fit of a computing technology to an organization's information strategy are often not asked, and yet this is the key issue. Computing is viewed from the standpoint of the hardware ("we've standardized on MS-DOS, the Macintosh, UNIX, the whatever") rather than the fit of a system into an information framework. Standardization on a particular type of computing technology (the computer itself, the network, etc.) is a partial step toward the real fundamental issue: does a particular purchase fit our organization's information strategy (assuming a strategy exists)? The intent of most operating systems or computer vendor standardization efforts usually comes in two forms: ease of support and the belief that using only one operating system will make sharing information easier. While these points are well taken, they don't ensure the desired outcome and may restrict or prohibit the use of key or new technologies. The important issue is whether or not the results of that system's use can be shared with those who need access to it. That issue is best addressed and satisfied through the development and application of an information strategy rather than policy decisions that say "we will only use x-brand of computers or operating systems".

THE NEED FOR A PLANNED APPROACH TO THE USE OF COMPUTING TECHNOLOGY

While the information needs of an organization change slowly over its life, the technology for working with that information is changing rapidly. The focus needs to be on the results of the work rather than the tools used since these tools will be changing on at least an annual basis, either by upgrades to existing tools (revisions to word processors, spreadsheets, data analysis systems) or their replacement as newer technologies emerge. At one point the state-of-the-art word processor stored all documents on magnetic cards. An office that had its letters, reports, and papers stored on that equipment would have had to re-enter those documents to take advantage of the recent developments in word-processing. (Later chapters of this book will look at current computing technologies and how they might affect your work in the future.)

The purpose of a strategic approach to the use of computing technology is to ensure that the products of your work don't loose their value as technologies change. The computer system you are using today will not be the same one you will use five years from now. Your strategy needs to assume that its hardware and software implementation will change over time. One of your goals for information management should be that changes in technology can occur without restricting your ability to work with the information and data already gathered.

HUMAN CONSIDERATIONS IN TECHNOLOGY PLANNING

Unless you are planning on a fully and completely automated facility, the effect of computing technology on people has to be foremost in your mind. The users of computers are people, and those tools are there to help them work. They need to be part of the process from the outset, something they participate in, rather than something that happens to them. Without their cooperation and support, the systems you implement—no matter how sophisticated—will fail.

There are at least three reasons for including the people in your organization in the process:

- They know how the work gets done, and their insight can help detect points of failure early in the process,
- Any changes made as a result of the process are going to affect them.
- Introducing computing technology can be a catalyst for raising the frustration level in organizations. Keeping people involved and informed may make the difference between a nightmare and a successful solution.

People are adaptable. In any system of rules, procedures, and regulations, we can overcome flaws in the process and deal with unexpected problems by making changes. Very frequently changes and adaptations are not put back into the system in any formal way; they just become part of the craft of doing a job. If you interviewed someone about the way they do a task, and then watch them doing it, you will find differences; differences that change a difficult or inefficient process into one that works smoothly, the "tricks-of-the-trade". One point of failure in "systematizing" a process is that the method used is the interview. It is quicker and easier on the designer than following people around. It also misses the short-cuts and "unconscious" procedures we follow in getting things done: the hallway conversations, the fixes to process problems that were never formally put into the corporate procedures manuals, the differences between how things are supposed to work and how they really work. Have the people who are going to use the systems critique the system's design and functionality to make sure it will accomplish its purpose. This will lengthen the design and evaluation phase and add some cost, but that added cost will be small in comparison to that of a system that doesn't function properly.

The New Work Environment

Computing technology is often referred to as a "change agent"—something that fosters change. It is fundamentally different from just being a better way of doing things, like substituting an electric drill for a bit-and-brace. Computing changes the work itself and replaces or removes tasks. That has a very

HUMAN CONSIDERATIONS IN TECHNOLOGY PLANNING

different effect on people than other forms of change like moving an office. Let me illustrate this with a personal experience.

About 15 years ago I was responsible for customizing and installing a system for instrument data collection. It acquired the data, did the processing, and printed a report with the results. One technician at that laboratory became very agitated when the system finally worked. Part of her job was to do the data analysis that the computer now did, and it was work she took pride in. From her point of view, a skill that helped define her value to the company and her self-esteem was now being performed by a machine. Where did that leave her?

In *The Age Of The Smart Machine* (Basic Books, 1988), Shoshana Zuboff details several such incidents; the book is well worth reading. One in particular is about the automation of a paper-pulp plant. Workers whose job was to keep the plant working based on their experience with the process (sounds, smells, the feel of materials) found themselves watching video screens after the plant was automated. The skills that they used to define their work, and so to an extent themselves, was no longer valued. They were replaced by automation and the new skills necessary to do their redefined work were difficult to incorporate into their hands-on view of the world. It was difficult to reconcile streams of numbers with the texture of a paper web.

The resistance to change shouldn't be underestimated. Nor should the benefits of careful planning and work toward obtaining user acceptance. A management article by Alice LaPlante (*Info World*, June 10th 1991, page S59) cites two examples. The first was a hospital conference room management system that was dropped because people didn't use it. The implementors saw the benefits in time savings, and ease of access, but enough people didn't see the point in changing to make the system successful. The implementors' view of change was never translated into benefits that the end-users could understand, nor did those workers agree with the need for change. A second example, a group manager convinced the users of the benefits of a Local Area Network for communications, replacing the manual transfer of documents. The implementation was not only successful, but fostered better teamwork.

The resistance to change can have several causes. Some may be technical, others reflect basic human behavior and needs. Implementing a system that allows someone to do "everything" from their computer could be viewed by management as a positive step. It would save time, promote efficiency, etc. It also ignores the point that people need to get up and move, stretch, talk to other people, and get out of their offices and cubicles. People need human contact. Using the telephone and sitting in front of a video display tube using electronic mail aren't particularly satisfying ways of making that contact.

One hurdle to using computer systems is the lack of familiarity with what they are, how they work, and the mechanical skills needed to use them—typing and working with a mouse or trackball. The language is different than people are used to, and the products mean a change in how work is done, even if the analogies between paper procedures and electronic ones seem obvious. To some, the fact that you aren't dealing with something tangible, like paper, make the results seem less real. The best solution to these issues is training courses. That training should be conducted by someone outside the organization, and, with separate sessions for different levels of personnel management. This stratification is needed for some very practical and realistic reasons. Different software packages will be used by different groups. The courses need to be tailored to a group's individual needs in order to avoid wasting their time (if they feel their time is not being used well or if the material isn't relevant, they won't come to the course). Senior managers may be reluctant to attend a "keyboarding" or typing course that is also open to more junior people. Acquiring new skills is an opportunity to make mistakes, and no one wants to appear foolish in front of anyone else. Very senior managers may want private sessions. Ignoring that point for the sake of efficiency or being practical isn't worth it. It just means that key people will not show up for the courses or learn to use the systems. If they don't find the systems useful, you won't get their support when you need to expand.

Most people under the age of 25 will have little problem working with computing systems. That also has its down side—they are already familiar with a system (maybe the one they have at home) and some software. That software may not be compatible with the systems you plan to implement or meet the goals that you will be setting. They will be faced with a change in their environment and may be reluctant to change. That reluctance may have a substantive basis or may be purely emotional (which doesn't make it any less real). Regardless of which it is, their adherence to what they have or are used to may take on a religious fervor.

One possible problem is that they are using XYZ word processor and you are considering standardizing on ABC. If they bring work home they will want to be able to easily work with that information. The simplest solution is to buy them a copy of ABC as long as the cost isn't excessive. If you are standardizing on a particular software package you may be able to purchase a site-license and be able to work home-use into that license.

The next problem that can turn the situation into a holy war is that they don't want to learn a new software package. This can begin as a feature-for-feature comparison. "Show me that ABC is better than XYZ"; "but I can do that with XYZ, why do I have to change?". Before you start choosing weapons, give this argument careful consideration. This won't be the last time you have to deal with it and diversification in software is going to become more and more common. Can the results of the work be transferred between packages easily? If so, is a hard-and-fast rule on choosing ABC valid, or is the real issue information interchange?

Twelfth-century Crusade style confrontations can occur when the computing hardware and operating system is considerably different, for example, the Macintosh and MS-DOS systems. Allegiance to a particular type of computer, particularly with those examples, is akin to religion. Before you mount your respective steeds, make sure that the issues are real. Is there a demonstrable

HUMAN CONSIDERATIONS IN TECHNOLOGY PLANNING

technical difference in the software package for a particular job? (Applications, which have the major impact on what you can do on small machines, are the real concern not the operating systems.) If the software is only available on one brand of those machines, there is little argument. Can the data be translated easily—test this carefully since some translations can loose or alter data. The general trend in computing is toward integrating and accommodating different computing styles—particularly on the part of software vendors. What may seem like a major point of departure today will be trivial tomorrow.

The implementation of a computing system can be seen as a replacement for valued skills, and, as the need for different and sometimes new skills. Part of the strategy development needs to address both these points so that people can see the change as improving their value and satisfaction with themselves and their work, and not as a threat.

Organization Impact

Introducing a computing system, or even a new computing application, can cause disquieting ripples in a small group. If that introduction affects how groups work together, things can get even more interesting. It forces people not only to look at how they do their work, but how they work with other organizations, and those relationships may not always be on the best footing to begin with. Managers will spend a fair amount of time explaining how their groups work and being surprised at the differences between their description and other managers' perceptions of what they did. It also puts these issues out in the open for anyone to see, including the next higher reporting level, and that may introduce some political energy into the process.

The process of developing goals may require the assistance of someone outside your organization or company. If nothing else than to act as a mitigating factor in spirited discussions and to keep the work on track.

Electronic Mail

The impact of electronic mail (EMAIL) will vary with the size and culture of the organization. Smaller groups, whose members are used to working directly with each other, will find it useful for communications outside the group: a faster substitute for hardcopy interoffice mail. Large organizations, particularly those in which people are under tight deadlines, will discover something else: the computer as a substitute for personal interaction. Companies that foster teamwork and have a free flow of information will be quicker to adopt EMAIL than those who have a strong, top-down approach to management and access to information. The latter group may view the easy exchange of information afforded by electronic messaging with disfavor.

The introduction of electronic mail into a work place can have adverse effects on people's behavior and interaction. Zuboff discusses the changes that took place in different companies when electronic systems were introduced. Most of the cases cited were in clerical positions at insurance companies that moved from paper-based systems to computer-based ones. In the chapter entitled "Office Technology as Exile and Integration", she cites the sense of isolation that people encountered as a computer workstation became the focus of the way they worked. They spent most of their time looking at a computer screen and much less time interacting with their peers.

The same set of problems applies to researchers, managers, and others as the computer becomes the focus of the way work is done; EMAIL just intensifies the problems that may arise. Among the changes that can take place are these:

- *EMAIL can replace human interaction, either direct contact or telephone.* How often have you gone to someone's office or called them and found they weren't there? EMAIL, and voice-mail, can be used to circumvent that problem by sending them a message. It will appear on their system, and they can respond via EMAIL, thus avoiding the "wasted time". Eventually this becomes the common mode of interaction—sending/receiving mail messages. People have less reason to talk directly to each other. You may find that the stranger sitting next to you at a meeting is the one you've been exchanging mail messages with for the past week.
- *Electronic junk mail.* The ease of using electronic mail is seductive. So is the use of distribution lists, it's as easy to send a message to a hundred people as it is to one. Because they happen to get on one list, and that list gets incorporated into another, people get mail they don't want or perhaps shouldn't get, or people who should get messages don't because they were left off the distribution list (a lot of message traffic can take place before this is discovered).
- *Changes in priorities.* EMAIL messages become the primary mode of communications and anything on "paper" is treated as a lower priority. If you want your message read, send it electronically. Anything else will be on the "to-be-read" stack because they can always "take it with them", "read it over lunch", or deal with it when hardcopy reading is the only reading that gets done.
- *Changes in people's understanding of what is going on around them.* You can create a document, send it to people electronically, and have a discussion about it via EMAIL, all without getting out of your chair. Then someone goes to your secretary and asks for a copy of that "hot" document and all the secretary can say is "what document?". Once the secretary was an intimate part of the work at the office, and now is outside much of what is going on. This reduces that person's view of his or her own value in a company.

This last point deserves some additional consideration. EMAIL can change

HUMAN CONSIDERATIONS IN TECHNOLOGY PLANNING

the functioning organization chart. In most groups there are at least two organizational charts: the one that exists on paper and depicts formal reporting relationships, and, the one that is practiced in day-to-day work. Human, face-to-face interactions are a lot more open than electronic ones even if you take into account typical office politics. Being on the "right" distribution list can have as much impact on getting work done as being invited to the "right" meeting. Not being aware of the information available on information systems, or not being able to get access to it becomes another piece of someone's feelings of frustration and potential isolation in a group.

While electronic mail can have significant benefits to a group, especially if they are located in different buildings, campuses, cities, and time zones, it can also have detrimental effects. Since the success of any implementation plan will depend on people's use of it, the human factors need to be carefully considered.

Environmental Concerns

The growing use of computing and networking technologies, in the office as well as the laboratory, has raised issues about the effects of various kinds of radiation on the health of workers. Those radiation types include:

- Radio frequency emissions and electromagnetic fields (including very low-, and, extremely low-frequency field) emitted by video displays. Pulsed low frequency fields (60 Hz) have been shown to be harmful.
- Microwave radiation from the growing use of wireless systems. Proposed systems use frequencies from 900 MHz to 18,000 MHz.

The effects of these emissions on humans are far from clear. Studies conducted in Canada and Sweden have demonstrated developmental abnormalities in chick and mouse embryos. Increased rates of cancer have been shown in children and power utility workers exposed to fields from high-tension wires. Recently concerns have been raised about the use of cellular telephones and the potential for brain tumors. This issue is far from settled on a scientific basis, let alone having been addressed by health regulations. Issues are raised in public forums with pronouncements that border on hysteria, just to get people acknowledge the problem. Independent studies raise concerns and lobby efforts raise questions about the validity of the studies. An overview-level article on the subject was published in 1990 (Paul Brodeur, *The Magnetic-Field Menace*, MacWorld, July 1990, pages 130–145) that reviews the then state of knowledge and provides measurements on monitor emissions. It is biased on the conservative side and does take manufacturers to task, but it is good introduction to the problem. (*Note*: there should also be concerns about the effects of emissions on laboratory instrumentation: are detectors and recorders affected and producing inaccurate readings, are signal noise levels being increased, are interfacing lines acting as antennas? Back to people.)

Rather than wait for a government requirement to generate action, your strategy should raise and answer these concerns. Shielding for equipment is available in the form of screens on video display terminals. Make them available either as a matter of routine or letting people know they can request them. In addition seating should be arranged so that each person is an arm's length away from the monitor and four feet from the next nearest monitor (this can cause a problem on typical desks, since they place the individual close to the system—pull-out keyboard shelfs can relieve this issue). Taking an active approach in minimizing the potential problems will make the lab personnel know that you are interested in them and that you are taking steps to eliminate potential problem; since people are among your most vital assets, their well-being should be accounted for. These steps may also be helpful in demonstrating due caution should court actions occur.

You should also make a decision about the introduction of wireless technologies into the lab. They represent an additional unknown factor on the behavior of equipment. You may be asked to demonstrate that they do not cause interference by regulatory agencies.

Carpal Tunnel Syndrome is a demonstrated problem in computer work. Carpal Tunnel Syndrome results from nerve damage with symptoms of numbness and loss of small motor skills. The problem can occur from carrying out the same hand motion repeatedly and by putting pressure on the bottom of the wrist for long periods of time—such as resting the wrist on a hard surface while typing or using a mouse/trackball. Proper training and the use of cushioned arms rest can minimize the problem, as can breaks and wrist exercises. There is a review of the problem including symptoms, causes, and means of prevention in the July/August 1993 issue of *Today's Chemist at Work* (pp. 42–44, published by the American Chemical Society). It's interesting that professional typists were trained to keep their wrists elevated above the typewriter keys to provide better control and more accurate typing—resting the wrists on the desk was wrong (see Figure 3-1).

Beyond that, there are other health concerns. In a typical, non–computer-oriented office, people tend to look for lots of light. Lighting may need to be changed when computer systems are introduced—the need will vary from individual to individual. Glare and eyestrain are the concerns. Some individuals will prefer to work in a more dimly lit office, others won't care initially but may develop a preference later on. There are anti-reflective screens that can be placed in front of a video screen to reduce glare, and they should be made available (some brands also reduce electromagnetic and radio emissions). Headaches from glare and eyestrain are real and can impair a person's ability to work.

In addition there are reports of nerve and joint ailments resulting from the use of keyboards. Concerns over these potential health problems are real and need to be addressed. While making provision for these concerns entails

HUMAN CONSIDERATIONS IN TECHNOLOGY PLANNING

Figure 3-1 Hands and wrists should be raised above keyboard or rest on a soft support, not on a hard surface or desk edge.

additional expense, as was stated earlier, the cost is well worth it. Computing systems that aren't used become very expensive items.

Voice Input and Output

One area that is getting a lot of attention is that of voice input and output (I/O) from computers. There are applications where voice I/O is useful. Transcribing recordings, doing inventory work, using voice output to help verify typed data, anytime when your hands are busy doing something else and a keyboard would get in the way are good uses for voice input. Voice and sound output is a more readily available technology, and it too has its place.

Macintosh users have had access to sound input and output for some time. Sound recordings (ranging from beeps to spoken phrases) can be attached to system events. Having a diskette ejected from a drive (a system command) can be accompanied by a variety of sounds, some of which are obnoxious. Some of these sounds can be useful. Getting an audio warning of an error rather than just having a message displayed on the screen will bring it to your attention even if you are looking at something else.

A product called the Voice Navigator is touted as a replacement for a lot of keyboard input, translating spoken commands into machine actions. These technologies can be very appropriate for handicapped users. Some see voice I/O as a replacement for keyboard input and a way of avoiding that issue altogether.

Whether this technology is appropriate for your office or laboratory is a real environmental concern. Most offices and laboratories tend to be quieter places. Introducing computing systems tends to reduce the tolerance for noise. Voice I/O is going to raise the general background noise level, and people may object to it.

The people issues may be the most challenging part of defining a strategic approach to computing.

AN APPROACH TO PLANNING AND IMPLEMENTING COMPUTING TECHNOLOGY

Figure 3-2 shows the essential elements of planning and implementing a strategic approach to corporate, laboratory, and, scientific computing.

There are three levels: the *Corporate Computing Strategy*, the *Implementation Plan*, and last, the selection of *Tools* to make that plan come to life. Before you take the view that "this is something that you might do in a *new* facility", or be concerned with "...but we already have all this equipment...", let me assure you that already having an investment in computing doesn't change the process nor does it presume that you will have to radically alter your computing hardware. The systems that may be in place (in most cases *are in place*) become a factor in the second and third levels when you look at how existing facilities fit into the picture.

The need to meet the requirements of regulatory agencies, system validation, and ISO 9000 is forcing a reconsideration of the computing environment. The computer system—originally viewed as a faster tool to do things—has become a central factor in the way business and work are conducted. Most companies could not survive today without this technology. As a result the corporate and departmental information structures are as much a reflection of the success and quality of those organizations as is any other measure. A serious reconsideration of how information systems and technology are used and implemented may be mandatory for a group to remain competitive.

Remember that the emphasis is on working with the three key products of an organization's existence: knowledge, information, and data. Today's hardware and software the *current* tools for doing the work. Over time they will change. One of the benefits of this approach is to help make sure that change will be manageable, predictable, and beneficial.

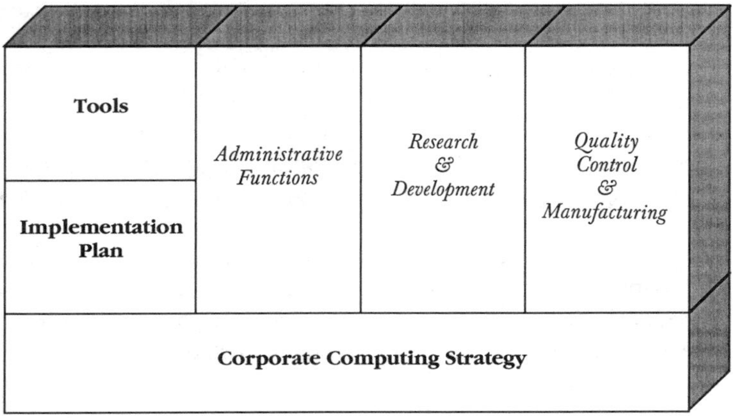

Figure 3-2 Essential elements of computing planning and implementation.

FORMING A CORPORATE COMPUTING STRATEGY

The *Corporate Computing Strategy* is the basis on which the implementation plan is designed and written, and ultimately, the tools and product selection are performed. The strategy is a response to a goal setting process that is described below. Note that it spans all groups that are part of an organization; there isn't a separate strategy session for each. The illustration shows departments that you would find in a commercial company, but the same logic holds for academic, consulting, and research institutions, *regardless of size*. The word "corporate" shouldn't be taken to imply that this approach is only appropriate for the private sector. It is important that this strategic approach be dealt with on an organizational basis. Not doing so will create conflicts when people want to share information, when systems are upgraded, or when new projects come on line.

The *Implementation Plan* is each group's response to the agreed-upon strategic direction. It doesn't have to be the same in each instance and should reflect each group's view of its needs. A department may choose to standardize on a particular operating system or a particular data management system, and its choices may be different from another's. The implementation plan should define the selection criteria that will be applied to products and where deviations are possible or under what circumstances deviations will be permitted. Its formation should include the opinions and concerns of the people who have to work with the equipment, in particular an understanding of their fears about any change that may take place, the benefits of those changes to them, and any training needed to make the systems useful. One characteristic of successful introductions of computing technology is that those using it *wanted* to. They saw the benefits to them, and they were trained and ready for it. This is particularly true of database systems. Those using one are going to be worried about what happens if they make a mistake. They should have access to the database system containing test data, so that they can practice, make mistakes, learn, try new things out without being concerned about the effects of errors. This "test" system may be around for a long time and provide a good test bed for system changes. No amount of computing technology will be of value if people don't use it.

The *Tools* level is the actual selection and implementation of the products and technology you've chosen to work with. This is where people get to touch and work with products, get trained, and begin to become productive with the systems.

FORMING A CORPORATE COMPUTING STRATEGY

Setting goals

Goal setting is the first step in the process. Those goals are going to reflect your particular organization and whatever is driving it. Clearly defined goals are needed to provide direction and as a measuring stick for the initial implementation and subsequent ones that will occur as change occurs.

Among the goals you might consider are

- maximizing the value and utility of data and information
- providing a computing environment in which change can occur in response to need and new technologies, with minimal impact on people's ability to work
- providing for a flexible work environment that supports people working remotely (the reason this is a goal, rather than an outgrowth of the system, is the need for systems security)
- securing systems against tampering or unauthorized access
- enabling people to work productively, without the computing systems getting in the way.

When you consider these and other goals there are three points to keep in mind: asset management, integration, and information flow. These points are considered next, and maximizing them may be additional goals to be considered.

Asset management

People who manage an investment portfolio are measured by its performance: a given investment yielded a certain return. The same holds true in real estate management and other dollars-and-cents oriented concerns. Information should be managed the same way. Money, effort, time, and people's energy are expended to capture data and information and it needs to be managed in a way that enables you to get the most benefit from the effort.

INFORMATION AS AN ASSET The following report has been repeated several times, the most recent telling was in the Colorado Gazette Telegraph (Colorado Springs, Jan. 2, 1991):

- "Two hundred reels of 17-year-old Public Health Service computer tapes were destroyed last year because no one could find out what the names and numbers on them meant."
- "The government's Agent Orange Task Force, asked to determine whether Vietnam soldiers were sickened by exposure to the herbicide, was unable to use the Pentagon computer tapes containing the date, site and size of every U.S. herbicide bombing during the war."
- "The most extensive record of Americans who served in World War II exists on the 1,600 reels of microfilm of computer punch cards. As the 50th anniversary of Pearl Harbor approaches, no manpower, money, or

FORMING A CORPORATE COMPUTING STRATEGY

machine is available to return the data to a computer so ordinary citizens could trace the war history of their relatives."

- "Census data from the 1960s and NASA's early scientific observations of the Earth and planets exist on thousands of reels of old tape. Some may have decomposed; others may fall apart if run through the balky equipment that survives from that era."

LABORATORY NOTEBOOK/DIARY EXAMPLE The above problems are not limited to government programs. Look around any R&D center. The standard data gathering tool is the laboratory notebook—paper. Over the course of an R&D group's life, people come and go, and filled notebooks are moved to library shelves. How hard would it be to find a particular piece of data or the results of a particular piece of work done by a colleague who isn't there anymore? Figure 3-3 shows, qualitatively, the cost/value relationship of the laboratory notebook. Since the notebook is essentially a technical diary, there are parallels in other kinds of work.

While the notebook is in use, both its cost and value increase. The cost reflects the effort and funding spent to get the material entered into the notebook. The value stems from two points: one is the value of the information, and the second is that while the notebook is in use, the scientist is well aware of the contents and that content can be put to use. Once the notebook is filled, the usual practice is to log it into the library, with the

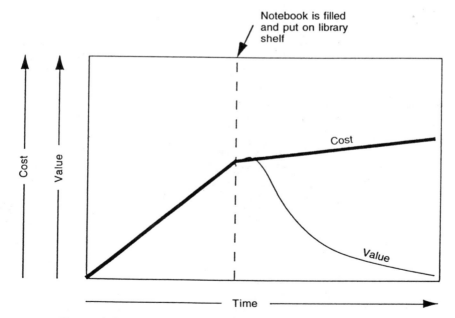

Figure 3-3 Cost/value relationship of laboratory notebook.

scientist keeping copies of needed material, and then to place it on the library shelf. The notebook retains its value as long as the material is accessible (someone hasn't moved it off into an archive because shelf space is at a premium) and the people who need access to the material recall that it is there, and what notebook it can be found in. As people and projects change, both the awareness of the material is reduced as is access—the value drops because the content is forgotten or it can't be located, not because the inherent value of the data is diminished. The cost continues to increase due to the cost of maintaining an archive. The key point here is that no matter how good the data is, it may be valueless because people either don't know it exists or can't find it. The inherent value or usefulness of the data itself may change over time.

When gasoline was 25 cents a gallon, not much attention was paid to research on synthetic fuels. As the economics of petroleum products changed, interest in that work increased and the value of the research did as well. Could the data be found when it was needed? Or was it a case of someone remembering that someone did the work "back then" and the data was in their notebook "somewhere in the library". What was the cost of doing the original work, what would be the cost of doing a search of the library for the data, what is the cost of duplicating the work if the data couldn't be found? How do you measure the value of the information against those costs?

While this is a hypothetical example, it does illustrate that information has a cost and value associated with it, and both those quantities vary with time, economic conditions, and need. If we treat information as an asset, how good a job is being done?

There are some other considerations to be taken into account, aside from the "process" issues. Lab notebooks are paper, and only a few copies of each exist (the original, the scientist's copy, and possibly a copy filed with the legal department). Paper has a finite lifetime and is subject to damage by fire, water, and various flora and fauna. Once those copies deteriorate the data is gone. Material can be put in controlled temperature/humidity cabinets, but there cost is a factor and someone is going to have to decide what to keep and what not to keep at some point. Microfilm shares the same issues, although the time span and storage costs *may* be more favorable.

"Computerization"—keeping all data in machine-readable form—is not, in and of itself, a solution. First, legal issues may require printed, signed copies. Second, there is a discipline in maintaining machine-readable files. In order for this to be a practical consideration there are some criteria that have to be met:

- *Backup.* The simplest is backup. If you only have one copy of the file on one disk, what will happen when the disk drive or the media fails? Power failures and voltage spikes are unkind to computers and disks. You really aren't in any better shape than if you kept it on paper. (I knew one individual whose approach to backup was to keep a second copy of the file on the same physical disk. This approach would solve a file corruption problem but that's about it.) Having a backup strategy is mandatory. That strategy should include not just one copy but at least two, and if possible on two different media; just in case you loose a backup copy because of a bad disk drive.

- *Cataloging.* The next is maintaining a catalog of the files and where they are located. Eventually disks, whether it is on a desktop computer or a main-frame, get filled either physically or are limited due to a budget constraint, and the data has to be moved to secondary storage.
- *Proper care and feeding of the media.* Disks and tapes are physical things, and the information on them is subject to chemistry and physics. Improper management of disks can cause data to be erased. This includes poor temperature and humidity control, and magnetic fields (there was a story about someone who kept an important disk right where he could see it, attached to the refrigerator with a magnet). Tapes suffer from "blocking" and data migration between layers in tightly wound reels. Disks and tapes should be periodically reread and copied to maintain the integrity of the information.
- *Keeping a copy of the files off-site.* In case of fire or other facility problems you don't want to discover that all of your efforts at backup and archive have been for naught because all of the media was kept together in one room and there was a fire, or the sprinkler system was accidentally turned on, or something else happened. Plan for the worst, it usually happens. If you are concerned about the cost, consider what the cost would be to recreate all of the information if it were lost. Would your group survive if all of the paper and machine-readable media at one location were lost? The 1989 earthquake in San Francisco and the effects of storms like hurricanes Hugo (1990) and Andrew (1992) cause some system managers to look at disaster recovery and institute procedures for protecting computers and the information they hold. Most of those efforts focused on large, central computer installations. Little, if anything was done for individual systems—they are usually left as the responsibility of the user who has little if any training in these matters.
- *Policy making.* Organization-wide policies and practices need to be instituted if storage and archiving of information in electronic form is going to be done. Some of these practices may become an annoyance to some, but then ask them how annoyed they'd be to loose a year's worth of research, or the final draft of a grant proposal. It is essential that the people who have to draft these policies and those that have to live with them work together. Policies are words on paper, if people don't understand them, agree to use them, and then do it, they have no value (aside from saying "I told you so . . . .").

Policies also have to take into account the fact that people who are working with information consider it to be "theirs" until they are ready to share it. It may be a simple issue of work being incomplete. In those cases the individuals have to be held accountable for properly backing up the files.

In its simplest sense, doing a backup is the process of making one or more copies of applications and data, usually to another disk or tape. If you have ever made a photocopy of a document "just for safe-keeping", you have done the equivalent of a backup. When you are working with a computer system there are considerations that can be more complicated. Photocopying a document for backup purposes has some built-in assumptions: you know what has to be copied, and the copier will likely have the ability to use the same size paper you are holding. Computer programs may use files that you aren't aware of, and the files produced can be bigger than the capacity of the diskettes built into the system. There are software packages that will assist you in doing a backup and they can work in different ways. Some will do a copy of everything that is on the disk which can be very time consuming. Others will permit an "incremental" backup; only making copies of new files or those that have been changed, and removing files that have been deleted. *Note*: the act of making a backup is only part of the issue. What you believe you have done is made a copy of everything on a system that is important. You can only be sure that is true if you try to "restore" the system from the backup copy (put a blank disk in and try to rebuild it). Backup systems may not always work the way you expect them to on all types of media. Always test the full backup/restore process to develop confidence in the process. About 17 years ago a customer asked if the system a colleague and I installed was properly backed-up and he could rely on it. We assured him it was and that he could. He promptly reformatted the disk (erasing everything on it) and said "prove it". Two things happened as a result: we proved that the backup process worked, and, we resolved not to do any more system installations on Friday nights (it was a three hour drive back to the office).

In order for a policy on backup to be effective, people have to be provided with the means of doing so, including both hardware and software. That will have an impact on budgets. It may be hard to convince a manager that a $2500 system (hard disk, floppy drives, monitor, keyboard, mouse, etc.) requires a $600 or $1000 backup device. The reality of the situation is that no one is likely to backup an 80 or 100 megabyte disk onto diskettes that hold a megabyte or less each and take up to an hour of someone's time. Even incremental backups take time away from more productive work. If you expect people to do daily backups, give them the facilities to do it easily. Provide backup devices, such as a removable hard disk or tape, that have the capacity to hold the entire system disk's data. That way the backup can be done in a single command or action that doesn't require their constant attention.

The "value" of information comes from its ability to be used. If you can't use it, because you can't find it, or it's in a form that hinders its use, or it's been lost, it looses its value.

FORMING A CORPORATE COMPUTING STRATEGY

Asset Management as a Function of How Work is Performed

The cost/value considerations also play a role in how work is done. The word processor is seen as something that improves "productivity", a given amount of typing and writing gets done faster and with better results using one than it would if a manual typewriter were used. The handheld calculator had the same cost/benefit/value relationship when compared to the slide rule. There is a point to looking at how those productivity gains were realized and how the value/cost ratio improved.

The process of typing a report with a manual typewriter was labor intensive and expensive. A draft of the report was prepared by the author, typed if the typist was lucky, but usually handwritten. The first step was a translation from written form to a typed rough draft.

In that step there was a net increase in the value of the information because the material could be more easily read (higher level of utility), understood, distributed, and revised. Each successive stage in the process of revision made some improvement in the quality of the information content but not the cost of working with it. Each typed draft required the same cost as the previous one, each having its own potential for new errors that had to be edited. It was only in the last stages that changes were few enough to be cumulative (using white-out to make corrections and then a good quality copier to hide the use of white-out—this couldn't be used in legal documents—or just retyping the pages that required it).

With word processors, all changes became cumulative. Once something was fixed it stayed fixed unless a content change was required. Reformatting of documents was a matter of changing the presentation and not the content.

One of the misconceptions that occurred was that the word processor was a device to be applied to the process of creating a paper document, and not as tool for working with information. Once the document was completed, it was frequently deleted because the job was done and diskettes were expensive. If a revision was required after it was deleted, the process began all over again; no gain in value, but considerable increase in cost. Paper is a way of distributing information and not the end product. The real end product was putting information in a more usable form. If it had been kept in machine-readable form, it could be re-used with little added cost.

Cost/value relationship change as a function of changing technology. The laboratory process of constructing a calibration curve from chromatographic or spectroscopic data is a good example. Prior to computers and programmable calculators, calibration curves were constructed by hand using a piece of graph paper. The data was plotted and the best "fit" was determined by a ruler and a sharp eye. The mathematical techniques for linear least squares fit were well known and would have given a more reliable fit, but the cost of doing the calculations by hand (with error checking) far outweighed the incremental value. The same held true for getting the measurements from the instrument. Peak area was more reliable for some chromatography than peak height, but the latter was used due to the ease of measurement. In spectroscopy, calculated absorbance from percent transmittance values was replaced with absorbance reading from specially ruled papers.

Once these calculations could be performed quickly using a machine to do the work, the more accurate procedures were put in place. The cost/value relationship shifted because the cost became acceptable for the increment in value.

Once you begin to look at information as something whose value can be affected by how it is stored and able to be used, you come to the next major consideration: how you actually work with it.

Integration

Your ability to work with information is a function of how it is stored and the tools you use. The word "information" is a very broad term, and as it's been used so far has included all forms of facts, figures, data, conclusions, and so on, that people work with. In the next part of the discussion, we're going to break "information" into three components: data, information, and knowledge.

The definitions of those three terms are difficult to phrase, and that difficulty is important because their definitions are very much dependent upon context. *Webster's New World Dictionary* (paperback version) defines "data" as facts or figures from which conclusions can be drawn, "information" can be data that is entered into a computer [definition 3]). "Knowledge" is defined as (1) the fact or state of knowing, (2) range of information or understanding, (3) what is known, learning, (4) the body of facts accumulated by mankind. Put together a committee that covers all the departments in an organization and ask them to define a "corporate database of information" and you'll get some idea of the diversity of opinion. Those efforts usually result in a compilation of sales and marketing "information", with technical "data" and "information" left to be resolved later. Rather than trying to resolve the problem in its global perspective, let's define the realm of concern to the stuff that is stored and worked with in a computer system.

"Data" (D) consists of individual measurements, numbers, or qualitative descriptions (the color of the bird is "yellow", etc.). "Information" (I) is the result of a process on data that creates or generalizes a relationship and can be represented by numbers, equations, graphs, drawings, and text. "Knowledge" (K) is a result of a process applied to "information" and/or "data", and is usually expressed in terms of graphics and text. It is the expression of K.I.D. that we will be concerned with here. Figure 3-4 shows the relationship of K.I.D. from a systems perspective. The ovals represent the storage of K.I.D. and the arrows are the processes used to work with one of the elements and produce another. The definitions of the processes are, again, context dependent, and not all of the processes may exist in a given work environment. The bottom arrow represents the data acquisition step in which "data" from the

FORMING A CORPORATE COMPUTING STRATEGY

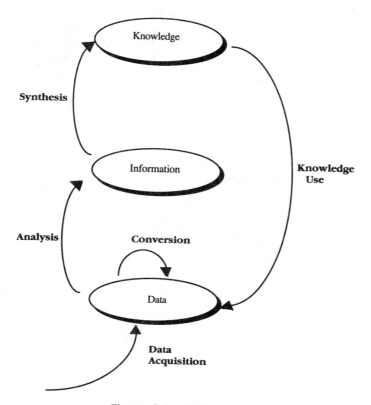

Figure 3-4 K.I.D. diagram.

outside world enters the system; it could represent an experiment or typing numbers and text. The arrow that loops back on data is a conversion step where data is changed from one format or representation to another, but no new data is entered (changing temperature from Fahrenheit to Centigrade for example). Data is analyzed to produce information. Knowledge is the result of the synthesis of information and can be used to create new data (through experiments or simulations).

The way the processes and storage are accomplished can vary widely. The entire paths for some work could be done with paper and pencil. Data acquisition could consist of writing measurements into a notebook, using a computer system to control an experiment, or just typing numbers into a spreadsheet. The analysis process could be manual or fully automated. The "use of knowledge" might be the skill needed to follow a procedure, a computer simulation of weather, or the mathematical model for an economic system.

A couple of examples will help illustrate the process—one from manufacturing and one from office work.

MOVING TOWARD A STRATEGIC VIEW OF COMPUTING TECHNOLOGY

One type of work in a quality control laboratory is to determine how much of a component is present in a material. It might be the amount of antioxidant (a compound that inhibits degradation of the polymer) in polyethylene or in a potato chip (check the list of ingredients on the bag the potato chips came in see if it has any BHT). The process involves the following steps (see Figure 3-5):

1. *Understanding* the instructions for carrying out the analysis method ("Knowledge" in this example includes the written method and the skills of the analyst),
2. *Preparing* the sample and getting the equipment ready to do the testing,
3. *Performing* the tests on the sample and standards (used to calibrate the instrument)—"data acquisition",
4. *Analyzing* the data to produce the analytical result—producing the information .

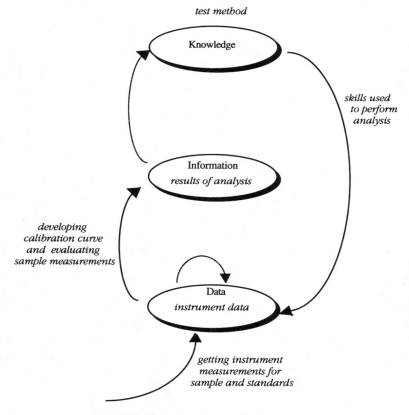

Figure 3-5 Using the K.I.D. diagram.

FORMING A CORPORATE COMPUTING STRATEGY

Note that in this process, both the *synthesis* and *conversion* processes are not used, though they could be. Synthesis could be the process of taking information (the assay results) over a period of time, performing statistical analysis on it and looking at that data and other manufacturing process information to draw some conclusions on the behavior of the production line.

In an office, the data acquisition step might be the entry of numerical data into a spreadsheet, where it is stored. That data could be processed to produce return on investment information, sales trends, and so on, all of which might be programmed into the spreadsheet.

While these may seem simple exercises, there are some serious implications for the design and implementation of computer systems, the choices you make in choosing tools, and the goals you may set in formulating your information strategy.

Integration is the process of putting a collection of "things" (hardware, software, databases, for example) together in a way in which they work together effectively. The diagrams above can be used to evaluate how well integrated a system is, and where the lack of integration is affecting the cost/value relationship of doing work.

Each oval is a place where the representation of knowledge, information, and data is stored. Each arrow is a process that acts on something stored. In a fully integrated system the input to a process should not require any manual intervention, nor should the storage of the results. Each process or arrow should have a functioning head and tail; functioning in the sense that input and output to and from machine-readable media are built in and the format of the input/output is compatible with other software. The office data flow is an example of an integrated environment because the data storage and processing were all handled by one program—this is the idea behind most integrated software packages; one or more programs each of which can access data from the other. Those kinds of situations are relatively rare in computing since most systems use software packages from more than one vendor and their data formats are not always compatible (work is being done to improve this situation). The quality control example is one in which the integration is likely to be minimal.

Data acquisition packages will have their own data formats, which are optimized for their particular problem. Statistical analysis packages have their expected input formats, which are sometime unique. The results of the analysis may be destined for paper, the screen on the computer, or you may have to create your own file output. What this means is that there are gaps in processing of data and information. Those gaps are usually filled by people doing typing. Taking the results of one program and typing them into a second is time consuming and costly. In many instances people have created data conversion programs that create intermediate data sets that have to be managed in addition to the original data.

The problem gets worse when you consider that those K.I.D. diagrams don't exist on a one-per-organization basis, or even a one-per-department basis, but that each person may have his or her own process for managing and storing those elements.

The cost/value considerations for managing "information" have to be extended to how it is actually used. Incompatible storage formats and the lack of integration provide barriers to making the full use of K.I.D., preventing you from realizing the full potential of research results and stored information. How much potentially useful work doesn't get done because it is too difficult or expensive to overcome the incompatibilities of data formats and programs?

Information Flow

Making information available to those who need it, when it is needed is an important part of asset management and integration. The knowledge, information, and, data we covered in the previous section are not the property of one individual but are usually shared by those within a group and between groups. Much of what is done in laboratory automation projects, scientific computing systems, and office automation efforts is heavily focused on the immediate group with little consideration for synergistic gains that could occur when groups take into account others who may need access to the same information or could benefit from a broader perspective.

People exchange K.I.D. in either a peer-to-peer or a hierarchical relationship (Figure 3-6). In a peer-to-peer exchange the classification of K.I.D. doesn't change as it would in a hierarchical situation.

An expanded version of the manufacturing example (Figure 3-7) is a good illustration of the process. Analysts in quality control exchange information about samples or tests in progress. The results of their work—information about the strength of a material, meeting color specifications, and so on—are transferred to the process control group. There is a change in context, from "information" to "data", that takes place as test results are fit into the overall data stream coming from other sources. That data is processed to give the engineers a picture of how the production process is working: their "information".

Frequently, process engineers will use EVOP (Evolutionary Operations Procedures) to make small changes in a process to see if it improves product yield, or some other parameter, and its effect on the economics of the plant. Combining information about the past history of the plant and the changes that are taking place gives them new "knowledge" about the behavior of the process. That "knowledge" ("as a result of doing X, yield increased 5%...") becomes a data point that the plant manager uses in reviewing the overall capabilities of the plant.

There are several points that come out this example. First, the definition of what constitutes K.I.D. varies from group to group. Second, because people use different kinds of K.I.D., expressed in different ways, their databases are going to be different. Third, the effectiveness of communications between groups, and the cost/value relationship of data, are interrelated (timely and

Peer-to-Peer Exchange

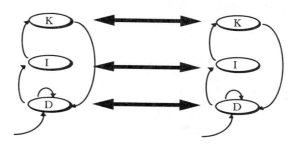

Hierarchical Exchange

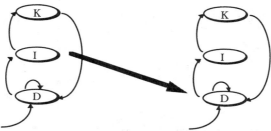

Figure 3-6 K.I.D. exchange relationships.

accurate exchange of QC information can have a large effect on a plant's operating efficiency, data arriving at process control too late becomes historical information rather than something that can affect the immediate process). Finally, the level of integration needed to gain full value from the K.I.D. needs to be planned for as an organizational goal, rather than an evolutionary process. These points have parallels in everything from academia to police work to banking, insurance, and other industries.

SUMMARY

The statement that "Computing technology is going to change the way companies function..." is only partially true. It *can* change the way an organization functions, but only if, like any other tool, it is used properly.

The first step is to put computing aside and look at your organization and what is going to make it function now, and five to ten years from now. The things that people carry with them through their lives are their experiences and what they've learned. Regardless of what your organization does, the

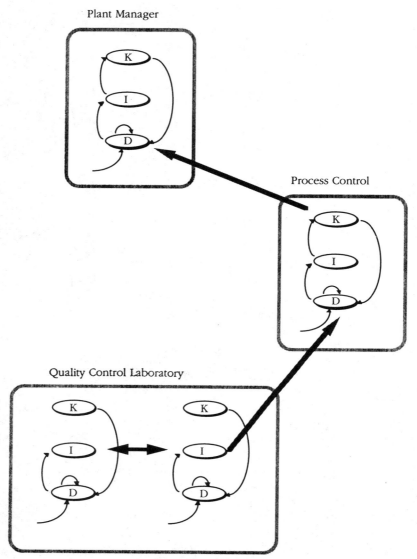

Figure 3-7 Manufacturing example.

things that it carries with it through its corporate life are knowledge, information, and data. Developing strategic goals for their management is essential to keeping the corporate organism functioning; whether that organism is a private company, university, government, or research firm it comes down to the same things.

SUMMARY

Once those goals have been established, the next step is to develop an implementation plan as part of a process to realize those goals. Our emphasis in the next chapter will be to develop guidelines for the evaluation of computing technologies as a tool for implementing that process.

It is important to keep one thing in mind: the goals you set should not presuppose that they are going to be implemented with computers, networks, etc., nor limited by today's technological constraints. The creation of new technology depends in part on a need for its development; your goals may just take a little longer to be reached. For the most part, computing technology is today's method of working with information. What will happen five years from now is anyone's guess, and flexibility is needed to take advantage of change.

CHAPTER 4

Developing an Implementation Plan

An information systems manager had been asked by the quality control director to look into a Laboratory Information Management System for their facility. The QC manager likened the purchase of the LIMS to buying a pair of shoes: try them on, if it doesn't fit, try a different pair. Analogies can be useful, and this one indicates that the manager didn't have a clue as to what the impact of a LIMS on lab is. During a discussion with the MIS manager, I suggested that the impending purchase was more like buying a home. The price scale is about the same, and so are the considerations.

When you look for a home, the physical structure is only part of the story. That part includes how it will meet your immediate needs and how well it can adapt to changes over the life of the family: using space differently as children are born, grow, and move out, adding accommodations for visiting relatives, and so on. You also have to take into account the neighborhood, schools, services, access to major roads, and other environmental concerns. The ability to meet current and future needs, plus access to services and facilities, plus price, are the major factors in buying or building a home.

The same type of issues are of concern in LIMS example. The structure is only the starting point. Adaptability, access to services, the ability to work with other data systems, and long term viability are major concerns. Once installed and validated, these are factors you'll have to live with for a long time. This kind of thing is not confined to LIMS, but applies to every data system considered for lab use.

The previous chapter gave us a start on specifying the broad criteria for the kind of "homes" we want our system to be. The material that follows will help describe the environment you work in, how you want it to function, the services needed to make it successful, and finally the specifications for the products that will make it real.

THE IMPLEMENTATION PLAN

The purpose of the implementation plan is to

- Determine the information flow and usage model that best describes how the group wants to work.
- Detail the interactions with other groups, and the ways in which a computer system can improve them.
- Define the criteria that will be used to evaluate and choose products.
- Determine the training requirements.
- Set priorities for the actual introduction of computing equipment in a new facility, or the changes that may be necessary in equipment in an existing computing environment.

Once you've defined the goals that your Corporate Computing Strategy is going to be based on, the next step is to begin working with individual departments to develop their implementation plans. These plans are going to be both a response to the goals that were set and an accounting of the particular needs of each group. The end result of this stage should be an understanding of needs and a set of criteria that can be used to evaluate products in use and those that will be needed in the future. *Note:* this is not "8 ways to a happier implementation plan"; the details of the plans will vary from group to group, and between organizations.

Each department or group needs to define the following points:

- the Knowledge, Information, and Data (K.I.D.) that it works with, who uses it and how it is used
- how frequently it is accessed
- if they could start from scratch, how would things be done differently (if at all)
- where does information and data come from, and what happens to the data and information that the group produces—linkages to other groups
- the critical bottlenecks in the information flow that affect people's work.

Part of the implementation plans' development involves a review of people's skills, and their concerns about changes in the way they work. Training programs may be needed to get people ready to work with new systems. One of the more successful installations of a major software system into a laboratory involved: (a) making sure that the people in the lab understood what was being done, why it was done, and what the benefits to them were, (b) understanding their concerns and reservations, and (c) training them to use the system. Part of that training involved the use of the software with a dummy database so that the analysts and technicians could try things out and make mistakes without being concerned with the consequences.

During this process you need to evaluate the different options open to you through different types of computing environments: personal computers, networked systems, time-sharing, etc. A description of various technologies is given later in this book. The key is not to pick and choose now, but to look at the strengths and weaknesses of the various alternatives and begin weighing them in the following discussion. The idea of "a computer on every desktop" is a current fashion (promoted heavily by those who sell those computers). Time-sharing is considered as "the way we used to do it". But there is much to be said for time-sharing systems: shared file access is easier, there is no network to worry about, and some applications (large database applications in particular) might be done better in a time-sharing system than in a network of PCs. The major drawbacks in time-sharing is the apparent large equipment cost relative to PCs (and the management level of approval needed to get one), the politics involved, usually poorer user-interface, and the need for support personnel. The cost issue needs be done in terms of cost-per-user, the cost of application development, and the cost of software license fees (it may be cheaper to have a multi-user license than a lot of single-user licenses; site-wide licenses are another possibility). The support-person cost is not always a real issue. Do you hire (or contract) someone specifically to provide support, or do you spread that cost out as a percentage of each user's time on a personal computer. Most users don't do a good job of managing systems on their own. Someone takes on the role of system guru and the time available for that person's primary work may be compromised—especially if the guru role is more satisfying and fun.

LOOKING AT THE MOVEMENT OF K.I.D.

The flow diagrams in the previous sections should be useful in documenting how data, information, and knowledge are stored and used. Each individual in your organization may need one or more to indicate how things progress. Different tasks may take different paths and have differing needs. This process is not going to be popular. Individuals have to ask and answer questions about their work and the results may be disquieting. Human nature being what it is, people may get defensive about the exercise ("Why are we doing this?" "What's going to change?"). Anytime an evaluation is done, anxiety levels are going to go up, so preparing people for it, and letting them know what is going to be done with the results is important ("no, you're not going to get fired, demoted...").

Among the things you want to look for are

- any place where the same K.I.D. is being manually entered into the

LOOKING AT THE MOVEMENT OF K.I.D.

"system" more than once (the "system" consists of all the paper files and computers in use in that department)
- situations where the printed results of one program's output are manually entered into another program, or copied from one form to another
- occasions where the same data or information is being stored in two different places (not counting backup storage)
- questions about where to find a piece of data ("If I need it, I just ask").

These are all instances where integration falls down, extra cost and effort are being built into the work, and where information isn't being well managed.

If data is being entered twice, manually, each instance is an opportunity for error, which means the data has to be verified. The same holds true for retyping printed results. If the same data is being stored in two different places, who worries about making sure both sets are updated if changes occur? If an individual is the group librarian, how big is your insurance policy on him or her? What would happen if that person weren't there anymore?

All of this comes right up against a major issue facing labs today: validation. The introduction of ISO 9000 into the regulatory mix puts everyone into the fray. Validation is a response to the challenges: How do you know your procedures work? How do you know the answer is correct? The answer to this challenge to a lab's workings is the summation of operating procedures, test method verification, equipment maintenance, software testing, and so on.

The points noted above face the validation issue squarely. If manual methods of data transfer are used, cross-checks of entries are required on a continual basis. If electronic means are employed, the transfer mechanism needs to be thoroughly tested, and only periodic checks need to be made to ensure that the process continues to function properly. While the primary purpose of this book is the development of a strategic approach to lab computing and automation, it generates a blueprint for that system's validation. The movement from general goals and principles to work flow diagrams, equipment selection criteria, and system implementation exactly parallels the requirements of an effective validation program.

Laboratory automation is a system design and engineering effort. Treating it as anything less is going to yield an expensive process of doing it over.

The "ideal" situation on paper may look like someone's depiction of "the perfect world". A nice idea, but how are you going to do that? Recognizing the difference between what currently exists and what you would like to have exist is a starting point.

The K.I.D. diagram of Figure 4-1 is the simplest form. It shows the K.I.D. flow for an individual or a group treated as an individual. There are other models, and the paragraphs below look at

- individual researchers working from a common database

DEVELOPING AN IMPLEMENTATION PLAN

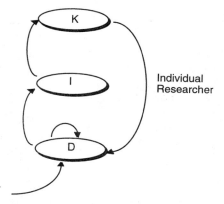

Figure 4-1 K.I.D. diagram.

- collaborating researchers working from separate "data" databases
- collaborations with a common source of data
- and two versions of a service laboratory (quality control, analytical chemistry, clinical laboratory, etc.)

Each of these forms has needs to be considered in applying information technology. This list is not exhaustive, but it may give you an idea of how to depict your own work or that of your group.

Researchers Working from a Common Database

The key consideration here is that all of the researchers (three in Figure 4.2, but it could be many more) have to gain access to the same "data". They may work in the same building or be dispersed over several countries. Some of the

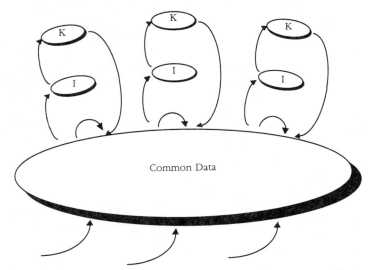

Figure 4-2 Individual researchers working from a common database.

large national research efforts in high-energy physics (Superconducting Super Collider), biology (Human Genome Project), and meteorology (Global Change) are examples of this kind of research. The goals here would be to ensure that each individual could find out if the data needed existed, get a *copy* of it, be able to use it, and enter new data in the system. In addition, you need to ensure the integrity of the data and provide protection against disasters.

The first two points, and the last, are issues of networking and storage management, and they will be treated in more detail later in this book. Finding out if data exists, aside from calling someone on the telephone, requires that a catalog exists. That catalog may be on paper (low cost, easy to duplicate, no system constraints, requires manual updating, indexing may be a problem) or available over a network (more expensive to access, some system constraints, should be easy to search, automatic updates from system librarian).

Getting a copy of the data is best done in machine-readable form unless you have access to optical character recognition services (human or computer). There are two choices for distribution in machine-readable form: copying over a network or distribution via media (tape or disk). Network copying requires that the user has access to the network and system (there are concerns over security issues) and a means of making the data transfer. Distribution via media requires that someone be there to do the copying and handle the administration of mailing and charges (someone has to pay for the material). Network copying has the advantage that you can get immediate access to data at any time, while media distribution is at the fate of the postal service and customs if national boundaries are crossed.

The third point, being able to use the data, is a bit more interesting. "Use" implies that you have the ability to read the data files, which requires you to understand that format of the data and be able to work with the data, preferably without conversion. Let me use some examples to illustrate the problem.

Suppose I had a sequence of outside-air temperature measurements that you needed and I sent them to you in the following form:

4	1	91	1	25
68	71	72	78	50
55	59	47	65	71
73	74	78	90	41
45	67	89	86	43
54	56	65	91	90

They wouldn't be much use to you unless you knew the starting date, the interval, and whether to read them column- or row-wise. That is one problem in data formating and organization. If you knew that the first line, or row, would mean that the first reading was taken April 1st, 1991, the interval is 1 day, and there are 25 readings, and that the rest of the data should be read row-wise, they'd be of more value. If you omitted the first line and gave the list to someone from China, they might take 90 for April Fool's day reading. If I chose to code the data so that letters represented numbers (A–J replacing 0–9), the data would be just as valid; it would just require more work to use it. The problem is more complex when computers enter the picture.

The Carbon Dioxide Information Analysis Center at Oak Ridge National Laboratory publishes a document entitled *TRENDS '90, A Compendium of Data on Global Change*. It is a printed document and well organized. You can look up atmospheric carbon dioxide, methane, and temperature measurements from sites all over the world. The printed format of the data is tabular and easy to understand. The machine-readable version of the data is available on diskettes for MS-DOS operating systems. Which is fine as long as you are using an MS-DOS system. If you are working with a Macintosh or large workstation system, you have to find a system that can physically accept the disk, and then use some conversion utilities to translate the files from an MS-DOS format to something your system can use. Since most researchers have access to MS-DOS systems and Macintoshes (with high density drives) can read MS-DOS disks, this isn't a major problem for most researchers. If you're a high school teacher with an Apple IIGS, it might be a different story.

The data files distributed by Oak Ridge are in ASCII format, the numbers can be read in as text and then converted to a numeric format that a spreadsheet could work with. Had they been encoded in binary format the situation would be more complex, since different computers interpret binary data in different ways. In short, the format of the data, the way it is encoded, and how it is distributed can limit its utility. If sharing data between users is important, as it is in the case we are dealing with, some planning has to be done to make sure it can be shared.

One of the first reactions people have had when faced with this problem is to "standardize", and the easiest way of standardizing is to assume or dictate some system constraints. The most common constraint is to set a policy that all people using the data need to have XYZ brand of computers. As the number of people working with the data grows, you will have less influence over the computing environment (type of computers used), and the less likely it is that this approach will be accepted. There are some other issues as well. One of the most serious is that choosing to standardize on a brand of computer, or even computers built on the same CPUs, ignores the fact that technology is changing. If your outlook is six months to a year, standardization on a CPU basis might suffice.

The Superconducting Super Collider in Texas was a case in point. Since this problem is currently in a state of flux, something may change by the time you read this. The high-energy physics people generate large volumes of data, all of which is stored in binary form to conserve space and reduce the time needed to work with the data (ASCII format requires a translation into numeric format, which can take a significant amount of time when you are working with large volumes of data). Their method of standardization (September 1990 through at least February 1991) was based on a binary format that was supported on *some* 32-bit workstations, not all. This limited their choices for computer vendors and did not take into account the fact that vendors were in the process of designing more advanced systems, which the physicists would be expected to want, that might be incompatible with their initial choice of standardization.

A better approach to the problem is to choose, or produce, a software standard that is not limited to a particular vendor's hardware. This method may have a penalty in performance, at least in the short term, but given the rate at which computer speeds are increasing, will not be a long term issue. It has the bonus of giving you flexibility in choices of computing environment and the ability to adapt to change. The high-energy physics group at CERN (Geneva) has created such a standard for their work, and similar efforts are underway in other disciplines.

While the examples used are from scientific and laboratory work, there are direct parallels in other situations. Sales data, police information, and most sources of data in the public domain fall into this category of work.

Collaborating Researchers Working from Separate "Data" Databases

In many respects this is the complement to the problem described above.

Here we have a number of people collaborating on a project, each with his or her own set of data from separate experiments (see Figure 4.3). The end result of the work could be a jointly authored report or publication. The basic difference between this and the previous example is the representation of "knowledge" (text and graphics); the issues of shared access are the same, although more emphasis is being placed on this area by commercial software vendors under the heading of groupware.

The basic idea behind groupware is that several people may be working on the same document, totally or in parts, at the same time. Each needs access to the current version so that they can see the revisions of their collaborators. Much of the initial emphasis is on textual material.

As was the case earlier, standardization is an important concern. While we're not concerned with the representation of numbers, we have just as complex a problem: word processors and graphics. Document preparation is essentially a problem in graphics, and the graphics entities can be divided into two sets: simple and complex.

On the simpler side we have the alphabetic, numeric, and special symbols that word processors and typewriters have been using for years. Each character can have different attributes (font, style, size, etc.) and be put together in documents with different rulers. Combining material from several sources into a single package is more than cut-and-paste when different word processors on different types of machines are used. The representations for

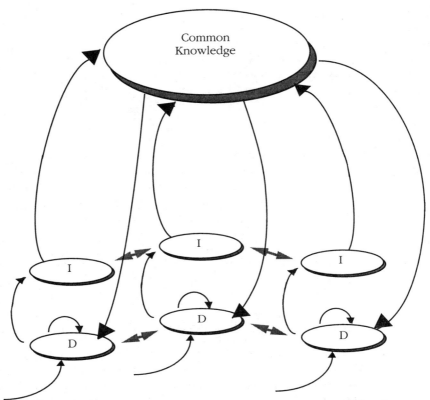

Figure 4-3 Collaborating researchers working from separate "Data" database.

attributes vary so that a portion of a document from one source may have to be translated into another format before the two can be combined. Some translation routines lose information so that a phrase that was bold in the original isn't in the translated version. The availability of fonts is another point. If one author uses a type font that isn't available on someone else's machine, changes will take place. Those changes may not be important if the entire document is in the same font, or it could be of considerable importance if the author used changes in fonts to indicate emphasis.

While most word processors have the ability to display a set of characters in plain, italic, and bold forms as well as others, so that a common workable subset can be agreed on, the same isn't true of more complex graphics used for illustrations. Different computer systems and graphics programs have different approaches to storing and displaying graphics images. Translation between graphics packages isn't as clean as you would like, even between different packages on the same machine; usually some information is lost due to features that exist in one program but not in another, because they are implemented differently or because the storage format that is used doesn't support all of the attributes of a given graphics system. If your work is heavily involved with graphics, you may want to consider one of several options:

- standardizing on the hardware/software package combination that ensures that people will be able to exchange and use images
- standardizing on a software package that runs on several different types of operating systems and computers and can import files from other systems
- standardizing on the format of the data underlying the graphics, so that the data can be transferred and different programs, on different systems, can produce images based on a common data set. This is best suited for graphs and charts.

The first option may limit your choice of vendors, but is the simplest to manage and implement. If you're careful about the program you choose (it's popular and the vendor is committed to support on several types of systems) that limitation may be short lived—it turns into the second bullet. The third is the most flexible for some types of graphics and for that reason the most desirable since it doesn't limit your choices.

In the area of standardization, text handling is a more tractable problem. The more popular word processing programs and page-layout programs run on many systems so you can adopt a group-wide standard and still leave the choice of computer system up to the user. One of the benefits of this standardization is that as new packages are developed and enhanced, the movement between them is resolved to dealing with a large number of documents in the same format, rather than the same number of documents in several different formats. This will be a consideration as library systems are developed and implemented.

Let's suppose you've built up a library of documents (your knowledge, base) and now you want to find all items that relate to a certain project. The electronic version of a document offers the potential for making this searching possible and quick. There are document searching systems for both mainframe and desktop computer systems that will assist in this process. BASISplus (Information Dimensions, Inc.) is one example.

BASISplus is supported on computers from several vendors including IBM, Digital Equipment Corporation, SUN Microsystems, Control Data, Unisys, Hewlett-Packard, and other UNIX (UNIX is a trademark of AT&T) platforms. The modules that come with the product allow you to import documents from different word processors into BASISplus's internal format, and then organize them, perform searches, and so on. It supports a client-server environment, so it is suited for a large networked facility.

The importing of information is, at this writing, a one-way street. Text that is brought in loses all of the character attributes (bolding, italics, etc.) that the original document may have had. You can move text out of BASISplus, but it may have to be reformatted. This means that you have to keep two copies of the document: one in the original word processor format and one in BASISplus. This is not necessarily a flaw in BASISplus, but rather a reflection on the lack of accepted and implemented industry-wide standards. Information Dimensions, the company supporting and distributing BASISplus, is looking at bidirectional conversion routines to at least mitigate the issue.

If your organization is looking to institute this type of system and you've standardized on one particular document format, it may be possible to build a specialized converter to move text in and out of the librarian. This is a much simpler and less costly process when only one document format is being used.

Collaborative projects assume shared work, and the facilities for it have to be designed in.

Both of the previous examples can be combined for collaborators working from common "data" sets (Figure 4-4). This is common for university groups who are part of large projects.

Service Laboratories

Service laboratories, including quality control, analytical chemistry, physical properties labs, clinical labs, etc., are worth discussing for a couple of reasons: there are a lot of them, and their functional behavior is typical of a number of organizations in and out of the chemical industry, so that a similar analysis can be applied to other groups. Figure 4-5 shows the information flow in a typical service lab.

The behavioral characteristics are these: individuals pick up samples to be processed, do the required analysis, report the results, and file the paperwork. In the most common situations, all of this is done using a paper-based management system.

The list of samples to be processed, the work in progress, and the procedures for carrying out the analysis are represented by the "Common Knowledge" oval. Each analyst is responsible for managing the data collected during a procedure and the results of the work, usually in a paper laboratory notebook (which shares all the problems noted in the previous chapter for managing the value of information). There are no arrows going from "information" to "knowledge" since these groups usually don't do test method development or independent research.

The issues that this system presents are these:

- The method of organization of data and information can vary between individuals.
- There is no systematic or common archive, aside from a paper-managing library, for collecting information.
- Searches for historical information require going through several docu-

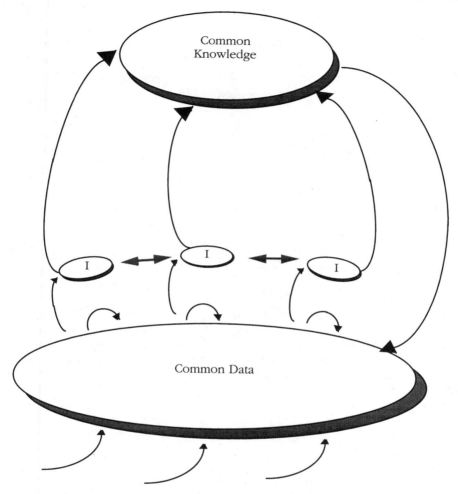

Figure 4-4 Collaboration from a common database.

ments, trying to find out who did the work, where the results are stored, etc.
- A potentially rich source of information and data for evaluating a product's or plant's performance is almost unusable.
- Movement toward a tighter integration between the service lab and its "customers" is likely to be frustrated.

A more effective K.I.D. flow is shown in Figure 4-6.

The major difference is that the result of each analyst's work resides only temporarily in a short term "data" database before it is integrated into a larger database of experimental data. This integration allows the development of a

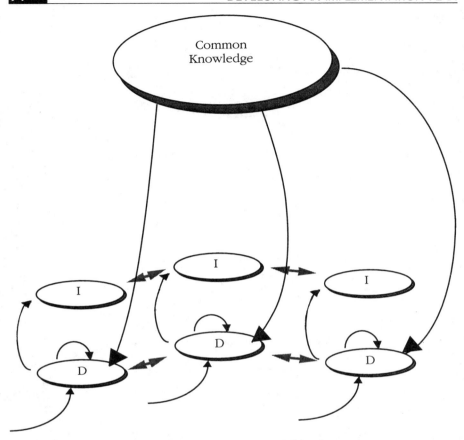

Figure 4-5 Information flow in a service laboratory (example, Quality Control).

data library or archive—the Data Librarian in Chapter 2 that makes searching feasible and provides a mechanism for the movement of data back and forth from long term archival storage.

The common "information" database makes better result tracking possible and provides the basis for tighter integration between the service lab on those who submit work.

While this is a nice picture on paper, its realization in a working laboratory is another matter. The "Common Information Base" is usually implemented as a LIMS. There are a number of them on the market from several vendors. Each addresses the basic issues of sample submission, tracking, results entry, reporting, etc.

The sticky part is the management of the experimental data. Most instrument vendors that either provide data systems or output to a data system (through a serial port or other interface) have their own method of specifying the data format. These formats are usually proprietary and have little

LOOKING AT THE MOVEMENT OF K.I.D.

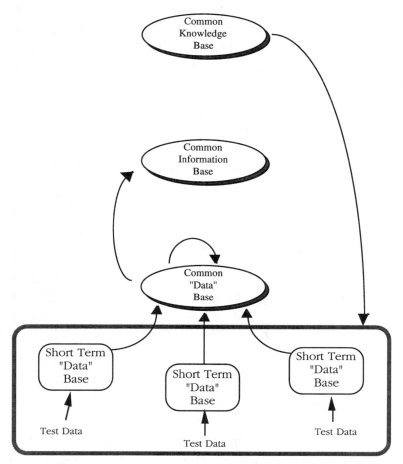

Figure 4-6 Alternative service laboratory information flow.

relationship between vendors, even for the same type of data. The development of that "Common Information Base" is going to be a user-specific effort. That effort should be well worthwhile since the standardization would permit data analysis routines to work on similar instrument data, regardless of the vendor of the instrument or data system. In addition, should an industry standard structure for this type of database be developed, then the translation between formats will be simpler.

Some steps in this direction have already been taken, though tentatively. The Analytical Instrument Association is working to develop standards for instrument data formats. The first effort is in chromatography, with mass spectroscopy and optical spectroscopy following. Should a broader range of standards result, and should they be followed and supported, then you have the start to a common database for experimental data.

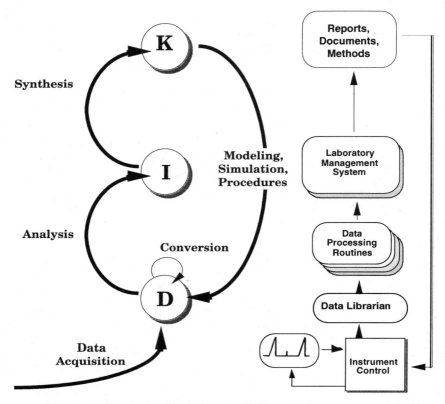

Figure 4-7 Relationship between K.I.D. and lab process models.

There is one problem, and that comes in the form of the Good Laboratory Practices as formulated by the Food and Drug Administration (this only applies to industries that are regulated by the FDA, although they are in general good guidelines for operating a laboratory). The issue is the definition of primary data. The instrument data stored in the vendors format would be considered primary data, while the converted formats may not be. This means that the instrument data still has to be maintained for reference and validation purposes. The Environmental Protection Agency has changed its stance on the definition of raw data. The Good Automated Laboratory Practices (GALP) has recently (Spring 1993) deferred the definition of raw data to individual EPA programs. As a result there is no uniform EPA standard for this data.

Figure 4-7 shows the relationship between the K.I.D. model and the lab process model of Chapter 2.

There are three sets of data bases:

- Reports, documents, and methods constitute the "knowledge" database
- Laboratory management system is the "information" repository,

COMMUNICATIONS BETWEEN GROUPS

- the data librarian contains the collected experimental "data"

The data processing routines are part of the analysis arrow. This arrow actually has a number of parallel components, one per test procedure or process. Sample preparation, etc., is part of the data acquisition step.

Interim Summary

Considering the material that's been covered, it's worthwhile to take a look at where we've been and why.

The preceding portion of this chapter has been concerned with the issue of K.I.D. flow within a group. How you use and organize these three elements is of fundamental importance to the success of any effort to use computing technology. Careful consideration of how K.I.D. *should* be stored and managed will have significant impact on your ability to take advantage of new technologies, optimize your use of equipment and people's time and talents, and extract the most value from your growing investment in these assets. It will also help to minimize the cost of their use.

Emphasis has been placed on standardization. First of all, standardization is not a goal, but rather a means of achieving a goal. It will help provide for consistent methods of access to information and avoid blind alleys in product and technology choices. Second, having a standard helps us recognize when a deviation occurs and gives us a means of measuring its impact and worth. Standards are guidelines and should not be viewed as rigid, unbreakable dictums. A carpenter's level tells us when something is out of plumb, but it doesn't tell us if it is wrong. It tells us that something is different from a "standard", and that exception may be worth the deviation. A new product may provide some benefits that exceed the inconvenience of introducing another storage format.

The purpose of the implementation plan is not to choose products, but rather to define the criteria that products need to meet before you consider them for purchase. Some of the benefits of this planning effort on your part will become apparent in two places: the next section where we look at communications between groups, and in the evaluation of emerging technologies.

COMMUNICATIONS BETWEEN GROUPS

In Chapter 3 some time was spent looking at the exchange and translation of K.I.D. between groups. Part of the value of your work depends on your ability to use what you produce, and part stems from the ability of others to use it—those others can be considered your "customers". Whether they are a peer group that you work with or an outside organization that is paying for a service, they are obtaining something from you and part of its evaluation is

DEVELOPING AN IMPLEMENTATION PLAN

going to include its utility. The utility of your "product" may vary with the "method of delivery".

The earlier example involved a process control example, and we'll stick with that scenario. It is a useful one since it involves feedback and clearly shows the relationship between the service group and its customer.

Figure 4.8 shows the process. Samples enter the QC lab for testing. The test results are manually transferred to process control, which uses those results to make changes to process conditions. If the time lag in testing is long or if there is a delay in using the test information, the test results become less useful because they may arrive too late to influence the process for a particular batch of material (this has nothing to do with the validity of the testing). Delays in getting results and reliance on manual updating of process

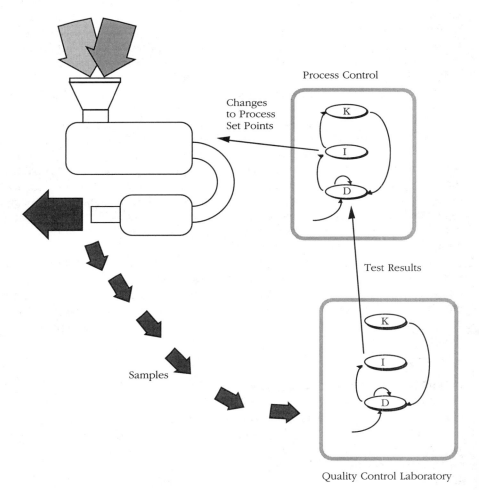

Figure 4-8 Process control example.

COMMUNICATIONS BETWEEN GROUPS

logs turn many test results into historical information rather than a useful part of the process control loop. High levels of automation within each group are of only limited value if the overall flow of data isn't affected.

When information flow is taken into account in the design of information systems, the movement of test results from the QC lab to process control can be done over a network or between processes in the same computer. The laboratory's information management system can feed the information to the supervisory process control system, which incorporates the data directly into its database. This results in a tighter integration between groups. The value of the test results is increased because the format is more usable.

The same type of thinking is necessary for research and office work. What is the best way to provide your customer with the results they need? For small amounts of information, paper reports may be appropriate and in some cases required by regulatory agencies. Once you get beyond that, transmitting a copy via diskette or over a network may be more effective since it minimizes transcription errors and makes the information available in a format that can be immediately used.

Written reports, particularly those that contain graphics, may still be necessary in the short term; they are portable and easy to work with. However, the developments in document management systems, and "hyperinformation"—which will be covered later in this book—are going to require electronic transfers.

Electronic Mail

There are two forms of electronic mail available today: FAX and computer generated mail. To some extent, because of the availability of FAX modems, the distinction between the two is getting blurred. The paragraphs immediately below refer to computer generated mail of the type you might see on a VAX, using Microsoft Mail, or available through an on-line computer service such as America Online, CompuServe, Genie, etc.

Electronic mail is one way adding a communications channel within and between groups. A brief discussion of it is included here since its use needs to be a planned activity and because of its impact on systems configurations and people. In its simplest definition, electronic mail (EMAIL) is the computer equivalent of paper mail (memos, letters, etc.). Some of what is stated below may seem simplistic at first, but they're worth saying.

In its most basic form EMAIL lets you exchange text messages between you and anyone else on the system or network; more sophisticated systems let you exchange graphics, sound, and video. That means that you are working—or planning to work—in either a time-shared environment or a networked system (if you are sharing a single-user system, use Post-It notes). The possibilities can be exciting, but

- It assumes that senders and receivers of messages can display the messages on their screens and (highly desirable) print them. If you recall what was said earlier about standards in graphics, this is not a trivial point—the kind of mail you want to send may put constraints on the hardware systems. Most people default to text messages, but there are still incompatibilities that need to be taken into account—the same points were covered earlier in the discussion of document standards.

- EMAIL usage in larger organizations can have a significant load impact on the network you use. This has to be figured in when you choose the type of network and its capacity.
- EMAIL can have a significant impact on system storage requirements. Many of the characteristics people show in managing their paper mail will be reflected in their use of the electronic version. If you are a pack-rat with paper mail, you probably will be with EMAIL. It is very convenient to save the electronic versions of mail—they don't clutter up your in-box—particularly if the EMAIL system has the ability to store message in folders ("To-be-done" is my personal favorite). Messages tend to stay around forever, occupying disk space. This can be a source of lively discussion between EMAIL users and system managers when it is time to "purge" accounts or move mail to off-line storage. Policies need to put in place early to avoid confrontations. System managers need to understand that people will always want access to that "important memo" that hasn't been touched in six months, and users need to understand that there are limits to storage capacity.

FAXes add an interesting dimension to the electronic mail issue. On the surface, they may seem outside our concerns because in most instances the result of FAX transmissions is the movement of paper (a piece of paper goes in here, another one comes out there). FAXes are very useful since anything that can be put on a piece of paper can be transmitted (in black-and-white). Problems can arise when people want to work with the information contained on the FAX.

FAX modems allow for the computer-to-computer transmission of text and graphics, as viewed by the human eye and mind. Viewed from the standpoint of the computer, it's a different matter. FAXes are a collection of dots and a FAX image in computer form can take 200K bytes of storage per 7.5 × 10 in. image scanned at 150 dots per inch (DPI). 150 DPI is a midrange resolution for FAXes. The impact on storage can be significant.

The next issue is actually working with the "information" on a FAX. If we print the image as a graphic or view it on a piece of paper, you and I can read the text. In order to get at the "text" in a FAX message into a computer-usable form (for example, in a word processor), the dots have to be converted to characters and that requires optical character recognition software (OCR) or a typist (OCR/human).

OCR packages have the ability to work from either stored images or with image scanners, so the problem is manageable assuming good image quality and a sophisticated OCR package. This will have an impact on systems cost, since good OCR packages are expensive and consume computing resources. There is also no guarantee that the FAX message sent and that interpreted by the OCR program are the same. Good OCR systems are only 95 to 99% accurate. That means that on average one character in every two lines could be wrong. The converted messages need to be validated. If the intent of transmission is to provide re-usable text, then file transfers are the way to go.

The purpose of electronic mail—in any form—is to exchange information (text, graphics, software) quickly. That exchange implies the ability to get and use the information. This in turn means that the users have to be able to address each other (requiring compatibility between types of networks), view the information (file structure and graphics compatibility), and manage the storage needed for a large messaging system.

Network Considerations

The use of networking is common today. The Macintosh comes with its LocalTalk network facilities built into every machine. Setting up a few machines for file transfers and interprocessor communications takes about ten minutes. The same capabilities can be had on IBM PC clones, although the lack of standardization and integration makes the job more difficult and expensive. These networks are not suitable for laboratory use in time-critical environments; they were designed around commercial applications where delays may not be catastrophic. In laboratory situations and in cases where the lab is electronically connected to process control in manufacturing, the requirements are more demanding.

Consider two applications running on the same computer (see Figure 4-9). For the sake of the discussion we'll call them LIMS and Process Control. In an error-free environment, almost any network system could be used to work this system. It's error detection, recovery, and contingency planning that make it interesting. The most obvious failure is the computer quitting. Everything

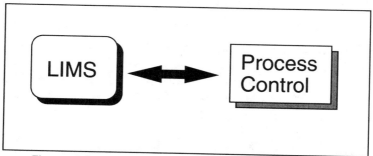

Figure 4-9 Applications running in the same computer.

stops, no communications take place. In a fully automated environment, what happens when the computer starts again?

The process control system has to

- Perform a self-check and make sure all of its components are operating properly (disks, files, etc.).
- Determine the current state of the process—a short interruption may not be significant if computer was the only element effected. If the disruption was long enough to cause process control problems or it was plant-wide, alarms would be set, safety precautions taken, systems restarted, etc.
- Find out if the LIMS application restarted successfully, find out what data was missing, re-establish contact, and update data and status information.

The LIMS system has to

- Perform a self-check and make sure all of its components are operating properly (disks, files, etc.).
- Re-establish communication with the process control application.
- Update that system.

Using multiple computers for the same situation significantly increases the complexity of the problem. That complexity stems from the number of things that can to wrong and the procedures needed to account for them.

The problems (see Figure 4-10) that need to be accounted for include

- *One or both systems becoming inaccessible.* Whether the cause was a system crash or a network failure, the surviving system has to be able to recover. That recovery should include automatically creating a journal file of messages sent but not acknowledged.
- *Both systems are working but the application is not available.* It may have halted, been superseded by a high priority problem, or be in the process of restarting. Messages sent may have been accepted by the receiving system but not the application. In that case the receiver has to journal the incoming messages, tell the sender that the target is unavailable, or both with failsafe corrective action taken. When the application restarts, it must check for any backlog of messages.
- *Disks may become filled or fail.* What corrective action is taken? Is there a secondary storage system? Does the system provide for disk shadowing (automatically making a copy of all data)? Can files span multiple disks?
- *Message or response is corrupted.* Is there a method of detecting corrupted messages and requesting a retransmission?
- *Application is being shut down.* This can be done for routine maintenance, update, backup, etc. Is there a graceful exit from the linkage?

COMMUNICATIONS BETWEEN GROUPS

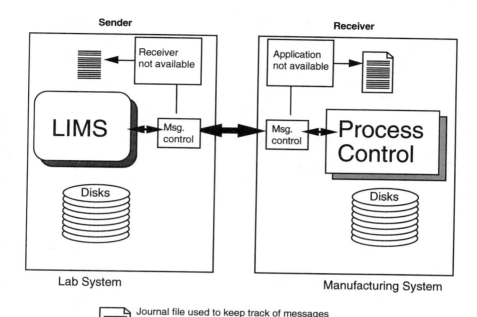

Figure 4-10 Applications running in multiple computers.

Many of these problems have been solved by older technologies such as DECnet (Digital Equipment Corporation) and the use of true multi-tasking operating systems. Systems designed for manufacturing environments may be better choices for laboratory operations than the more "modern" operating environments that have yet to encounter or solve these issues. The problem isn't the age of the operating system, it's the maturity, and "mature" as it is used here is an asset not a liability (try substituting "proven" for "mature" and see if your perspective changes). Plan for the worst-case situation, it always occurs—usually during a demonstration to your boss's manager.

The same concerns occur in research work. The use of robotics systems coupled with a computer for data acquisition has to deal with the same problems—looking at all the possible points of failure and taking them into account in the software on both sides of the link. One example of the use of a robot in this type of work is illustrated in an article in the May 1993 issue of *Chemometric and Intelligent Laboratory Systems* (C. Driscoll, J. DePaolis, and A. Vorbrodt, "Automated flexural testing: coupling of a robot-assisted system to a minicomputer", *Chemometric and Intelligent Laboratory Systems: Laboratory Information Management*, Vol. 21 #1, May 1993, pp. 75–84). In this publication, a combination of a robot, tensile tester, and supervisory computer is used to conduct a physical property measurement on materials, provide automated data collection and analysis, and log results. Among the points that need to be considered, in addition to those noted above, are any safety factors and system control parameters. For example on the robotics side: shutting down any moving parts and heating/cooling equipment that might lead to an unsafe condition and notifying an operator, through external alarms, that a problem exists and corrective action needs to be taken.

SETTING CRITERIA FOR EVALUATING PRODUCTS AND TECHNOLOGIES

The previous sections reviewed some points that need to be considered in formulating an implementation plan: the type of organization you have, the need to exchange material with peers and other groups, and the general issues involved in communications via electronic mail. With that background, the next step is set criteria for the actual products that will be used in bringing that plan into practice.

Those products are going to consist of hardware and software.

So far we've been considered the K.I.D. model as a two-dimensional structure. It does have a third dimension, which depicts the layers of hardware and software needed to implement a function.

Figure 4-11 shows the databases consisting of: a database (DB) access manager, the database software itself, the operating system (O/S), and the hardware. The application arrow contains the application, supporting libraries of functions (Dynamic Linking Libraries in a Windows environment for example), the operating system, and the computing hardware. If the two databases drawn here and the application all resided on the same computer, then the application would access the databases directly. If they were on different computers, then network access (the bidirectional arrows) would come into play through system calls to the O/S, then the hardware, etc.

One of the points raised in the section on setting goals was the need for integration between software systems. In order for real integration to occur an

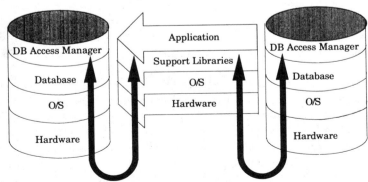

Figure 4-11 Databases.

SETTING CRITERIA FOR EVALUATING PRODUCTS AND TECHNOLOGIES 85

application needs to be able to read from the data source directly (without intermediate conversion steps), and then write to the destination database. In Figure 4-11 the application software, through support libraries and the DB access manager, would gain access to the necessary input data and, when its work was done, produce output. This ability depends on standardized data structures and access routines.

Without that integration capability, you have to rely on either manual methods of data input—with the inherent cost of validating the input—or conversion software, which adds additional database–application–database linkages into the systems. Each one of those linkages compounds the software development process and the cost and effort of validating systems. Your implementation plan needs to test for integration, and wherever possible, rely on electronic data transfers.

Among the considerations in specifying criteria for products should be

- the functions that the products are to perform
- their ability to read directly from the data source
- their ability to write directly to the results destination

Integration can be carried too far in regulated laboratories. Both Macintosh System 7.0 and Microsoft Windows 3.1 (and later versions in both cases) have variants of an automatic update capability. In the Mac system it is called Publish and Subscribe; in Windows it is Dynamic Linking and Embedding. Other operating system environments have similar capabilities; VMS workstation systems have "hot links" for example. These facilities allow the user to have different software packages automatically update documents when a change occurs in an underlying data structure. This is nice in commercial/office applications, but deadly in laboratory systems because of the possibility of having data unintentionally changed. These features can be selectively enabled and disabled, but people have to be made aware when their use is appropriate and when it should be avoided.

The rationale for these features, particularly in commercial applications such as presentations, runs like this: you have a presentation consisting of a number of slides, some text, some drawings, and a few graphs. The data that the graphs is based on changes in a spreadsheet program. Through these linkages, your charts are automatically updated instead of requiring you to delete and recreate the illustrations. This is a nice facility for charts and useful in a number of applications. It also violates every principle that GLPs, cGMPs, GALPs, and ISO 9000 revision control is based on. One requirement of the FDA/EPA regulations is that any changes made to data be done in a orderly manner. The original data cannot be deleted, and changes have to be checked, the reason for the change noted as well as who made the alteration. Hot Links, Publish & Subscribe, Object Linking, and other similar technologies do not provide support for the audit trails required by these agencies.

Software and hardware technologies introduced into the laboratory have to be evaluated not only on the basis of functionality, but also in their ability to help the lab meet data management goals and regulatory requirements.

PLANNING FOR THE INTRODUCTION OF REVISED/UPDATE SOFTWARE

The commercial software industry is highly competitive and in a constant state of flux. Vendors' responses to competitive pressure are to add more features to products at what seems to be an accelerated pace. This keeps the cost of products down and the functionality high. It also put the results of your work in potential jeopardy.

The battle over operating systems is just one highly visible example of the problem. Operating system developers seem to be on a crusade to cram as much capability into their packages as they can. The most recent example is Microsoft's release of DOS V6.0. Of particular interest is the disk-capacity doubling software. It allows your hard disk to appear to have more data storage capacity by using lossless file compression routines to make more efficient use of disk space at the price of slower system performance. This technology was originally used for archiving files in compressed form to reduce the size of the archive and to reduce data transmission time between systems over a network (files were compressed to about half their size, sent, and then expanded for use—the compression/expansion times were small compared to potential savings in transmission time). Aladdin's Stuffit and PK wares PKZip are other examples of the same capability, and there are more.

Earlier implementations of file compression software were done on a file-by-file basis, giving you time to verify to your own satisfaction that the compression did not loose data. The implementation in DOS 6.0 does the compression on-the-fly, so that a noncompressed copy no longer existed (users previously had the choice over what files could and could not be compressed). Due to flaws in the implementation (reported in *InfoWorld* in May and June of 1993) it was possible to loose data. In fact, you could loose an entire disk's worth of programs and data files.

Your implementation plan needs to have defined procedures for the acceptance and introduction of new software and revisions to existing packages. If you are in a regulated environment, this is mandatory and should be part of every validation plan. New features need to be tested and verified—not just to the extent that they work, but also to demonstrate that they do not lose older features' functions. Part of that plan should be the retention of older software versions. There is always the possibility that an "undocumented feature" may alter the way a numerical calculation is performed and give different results in different versions of the same package.

PLANNING FOR THE INTRODUCTION OF REVISED/UPDATE SOFTWARE

This is going to delay the introduction of new software technology into laboratory work, which will frustrate people ("my friend has this great software package but I can't use it because of the backlog in the group validation"). That is a very real issue; people will always want the latest and greatest of anything. Both they and those responsible for testing and validating software must realize that this need plus the need to make sure that their data is protected from loss, corruption, and alteration are satisfied.

An article ("Software upgrades are driving managers mad" by Doug Van Kirk), *InfoWorld*, Vol. 15, #29, July 19th, 1993, p. 58) offers the following list of suggestions for managing the upgrade process (added notes are in italics):

- Get on the beta list for software. That way you'll know what's coming and be better able to deal with it. *A beta list is the list of individuals who will test prerelease versions of software. This gives you experience with the product—and it prerelease bugs—before it is shipped.*

- Develop specific criteria for evaluating an upgrade. Although it's difficult to measure productivity, consider skipping upgrades that don't offer major advances in functionality.

- Skip the x.0 versions and wait for the bug fixes that inevitably follow in the x.0a and x.0b editions. *Vendors identify major new releases of software by version numbers. Those with version numbers ending in 0 are significant releases and those ending with another digit or letter are revisions. Version 5.1 is a revision of 5.0. Revisions are usually the result of bug fixes—sometimes significant ones.*

- Run software from a server. It won't be as fast, but you will only need to install it once. *A server is a separate computer used to store information (in this case). The user gets information and programs from the server and runs them on his/her local computer. This means that everyone runs a copy of the same software. The comment on "slower" is due to the need to load software over a network instead of the local hard disk.*

- Call the software publisher. Many retailers and distributors aren't aware of the full range of licensing agreements available from most vendors. *This addresses the cost of upgrades—site-wide licenses are often less expensive per copy than individual user-by-user upgrades.*

- Determine who needs an upgrade. If file compatibility is not an issue, you can put some users on a new version.

- Don't install software that has significantly higher hardware requirements on existing systems unless they're up to it. Slow performance may make some users abandon the program altogether or not use many features. *Software developers seem to assume that you have unlimited disk, memory, and funds. Upgrades mean new features, and new features mean more software and heavier use of resources. The hidden cost of a software upgrade is often the added hardware needed to support it.*

MAKE OR BUY DECISIONS

The decision to purchase a hardware/software/instrument system or build it from scratch has been shifting steadily toward the "buy" decision. There are a number of reasons for it. First, we now have things that we can actually buy that work in laboratories. Fifteen to twenty years ago if you wanted a laboratory system, you designed it and built it. Today most laboratory problems have been addressed by commercially available systems, either as turn-key products or as well-designed foundations for designing complete systems.

The former category includes instrument specific data acquisition systems, word processors, and molecular modeling systems. That latter consists of products like LabView, Labtech Notebook, Zymarks Zymate systems, Hewlett-Packard Orca robots, and others.

Building a product needs to be approached with a lot of caution. It may be the only option to solve a particular problem, but skepticism is a worthwhile attitude. Projects today need to be better documented, designed, tested, and supported. The economic climate requires careful consideration of how funds are used and how soon a return on investment can be realized. The regulatory and inspection activities of agencies pay particular attention to custom-designed systems: Are proper engineering methodologies being used? Is the code well documented? Has an independent audit been made of the work? How was it tested? Who tested it? What made the determination to build rather than buy? How well trained were the people doing the work? Does the design of the system mesh with the rest of the lab's data handling needs? Can the people assigned to the project manage a project, including having the ability and authority to say no to requests to add new features once the design has been agreed upon?

Projects tend to take longer and cost more than expected. Whatever estimate is given for time spent and money required, at least double it. It's not that people are dumb about scheduling, they're just human. They want a project to succeed so they're a bit optimistic on the skills and time required. The whole idea of a hard and fast project schedule is a bit unrealistic. If it hasn't been done before, how can you possibly know how long it will take to accomplish a task? Gut feel is fine, but remember that it is a guess and not reality until it is finished. Target dates and goals are useful for measuring the progress of work, determining if real progress is being made, and exposing serious problems. However, you shouldn't take those dates seriously until the 90% of the project is completed. Project management books and courses are readily available, so we can skip that subject.

There is also a support issue. The decision to build a custom laboratory system is also a commitment to support it. That support includes training new people, fixing problems, adding new features as they become needed, resolving conflicts with newly added software, and making sure that the software can function properly on new version of an operating system.

MAKE OR BUY DECISIONS

The real concerns come down to justification:

- What will the project accomplish?
- What will happen if it doesn't get done on time?
- How will the lab operate in the interim?
- Are there commercial products under development that come close to doing what you want to do? Or close enough that they can be augmented? Whatever you build, you will have to support for a long time.
- How much will it cost to build, document, test, and validate the project?
- Is it worth it?

The answer may very well be yes to the last question. If it is, make a big sign explaining what you are doing, why, and who is going to benefit by it. Six months after the project starts, the reminder may help keep a sense of perspective.

As much as possible, avoid building something from scratch. Take something that exists and add to it. If what you are building is valuable enough, maybe the vendor will cooperate with you and either offset the cost or add expertise in return for using the idea in the product. You will loose some uniqueness, but is that uniqueness what your lab's business is built on?

Should you decide to get involved in a project, please consider the following comments made by Dr. R.D. McDowall at the 1992 LIMS Conference (Pittsburgh PA), and expanded in a 1993 article (R.D. McDowall, "The evaluation and management of risk during a laboratory information management system or laboratory automation project", *Chemometrics and Intelligent Laboratory Systems: Laboratory Information Management*, 21, 1993, pp 1–19)—with editorial comments added:

The success of a laboratory project depends on

- *Commitment to the project by management*—that commitment includes a willingness to see the project through even when schedules slip (and they will) and to holding people to providing the necessary support.
- *Cooperation and support of the people that will use or be affected by the system*—that includes continued enthusiasm, prodding the developers to get the job done, and responding to request for information, comments, etc. Nothing kills a project quicker than indifference to its outcome.
- *Willingness to make decisions when needed*—some *person* needs to take responsibility for making decisions in a timely fashion and living with the results. Project costs increase and schedules slip if committees are left to form a consensus on every point.

Common causes of project failures are

- *Failure to learn*—from previous implementations or others doing similar work. Make sure people have taken the time to look at what has been done before, what worked and what didn't and avoid being entrapped by egos.
- *Failure to anticipate*—not taking into account the requirements of other affected organizations. Don't do system design in a vacuum.
- *Failure to adapt*—people not taking advantage of the opportunity for change, fear of new methods. Actively promote the benefits of the new techniques. Pick the individual most reluctant to want the new software/hardware/instrument and use his or her enthusiasm as a gauge for the success of your efforts. You will have to sell people on some ideas; they won't see things as "obvious" as you do, so sell them, literally. "Taking advantage of new opportunities" isn't restricted to whatever you are building. It is also a chance to review how things are currently done, how they could be done better, and make the change. Get people to view change as normal, desirable, expected. It is. It will either happen to your or with you.
- *Catastrophic Failures*—are the result of poor technology choices, incompatibilities with other systems, and poor project management. Make your choices based on their merit and long term prospects, not ease of implementation. Money is a poor basis for choosing the right answer. If you haven't managed a project before, take a course and learn how.

There is no substitute for doing it right the first time. Avoid the "there is never time to do it right, but always time to do it over" approach. Sacrifice schedule for the right answer, but be wary of too many schedule sacrifices. Sometimes you have to sacrifice the manager (or the developer).

One question that comes up in the decision is the justification for a project. Laboratory automation projects will rarely reduce existing headcount. Some, such as the introduction of LIMS, will temporarily increase the workload because of the need for running parallel systems during testing. The gains from automation come from reducing the long term need to add people and from moving people from routine mechanical tasks to those where human intelligence is better applied and improving the quality of data (during a discussion on this subject at a Good Laboratory Practices Workshop at the 1993 AOAC meeting, it was pointed out the most of the variability in analytical data comes from human sources).

TESTING THE IMPLEMENTATION PLAN

This has to be an egoless process. If it isn't, by the time it's over you may not have one anyhow. This is the most critical step in developing an implementation plan, and the first major outcome of the validation plan. Without adequate and widespread review of your direction, you may take a wrong turn.

Each individual who works in the lab, submits material for testing or is a "client" of the lab should have the ability to comment—the "outsiders" will be affected by the implementation of automated as well as those working in the lab. The strategy/implementation planning process we've described is very much like the process that software developers use in determining the functions and data flow path in a product. An examination process, similar to a code walkthrough, is appropriate.

During a code walkthrough, each step of a program is examined. The reasons why certain choices were made and other alternative discarded should be detailed. This does not have to be an adversarial process, but a cooperative one yielding an honest appraisal of the work done. You're better off finding out about a missing piece of information or false assumption here than after a lot of effort has been put into make it actually work.

Changes in the plan are likely, and not all suggestions have to be implemented. The choices as to which are and which are not should be documented. The end result will be a plan that has a better chance of success and more support by those in the lab and those in other departments. Any temporary upheaval in the lab's operation will be better understood, tolerated, and appreciated.

This approach is more difficult to follow in a research facility than in a testing laboratory. Testing labs are more stable in the operations, research is more dynamic and subject to change as work progresses. However, provision for change can be planned. Databases should have fields added that allow for expansion without re-creating the software. Standardized data formats should be used as much as possible to permit flexibility in moving and using information; if standards don't exist, establish one for work in your facility that will make future software and data handling problems easier to solve. Having done this, the walkthrough will be more useful, and the results more valuable to the successful operation of automated systems.

OF PARTICULAR INTEREST

Review of FDAS-483 Citations

Ty Fujiwara of the Food and Drug Administration prepared a memo dated May 18th, 1992, which reviewed the basis of FDA-483 citations between June 20th, 1979, and March 31st, 1992 (no further work has been reported since then). Of the citations, only 5% were for computer related issues (see Figure 4-12).

One of the recommendations from that study was that inspectors pay more attention to computer systems than they had in the recent past. From the reports we've been getting at the LASF, that is the case.

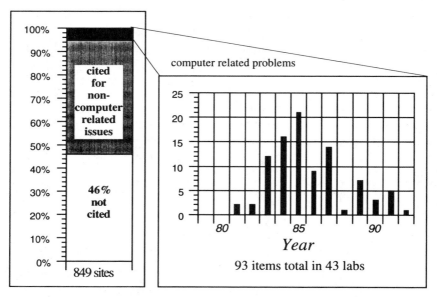

Figure 4-12 FDA-483 citations.

Figure 4-13 shows the distribution of these problems in four key areas. These are points that should be reviewed in your implementation plan—if others have missed these points, you may have as well. The Good Automated Laboratory Automation Guidelines from the Environmental Protection Agency, while not officially endorsed by the FDA, do provide a good basis for reviewing a labs procedures and further testing the basic design of an automated facility.

Standard Operating Procedures

While SOPs are needed in each phase of a laboratory's work, your implementation plan needs to address

- system security
- verification of data entry
- system backup and recovery procedures
- equipment maintenance
- software and system development
- system maintenance
- system validation

The material following is an introduction, and not intended to be a thorough

OF PARTICULAR INTEREST 93

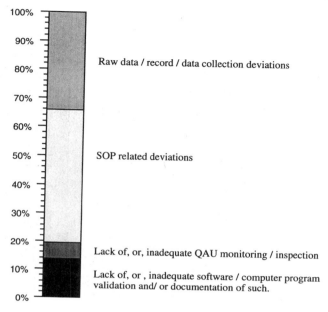

Figure 4-13 Distribution of computer-related problems.

treatment of the subject. There are texts, courses, and consultants that cover the material quite well.

SYSTEM SECURITY System security is a particularly knotty issue. It directly addresses the goal of asset management; making sure that K.I.D.'s maintain their value. That includes protecting them against unauthorized access and loss. The intent is not to provide a thorough treatment of each subject, but rather to introduce them so that they are taken into account in your planning.

Security and easy to use/access are divergent paths. Unauthorized access, preventing data alteration, theft, and password protection schemes are points that immediately come to mind when security is mentioned. Just as important are procedures that ensure a systems operation, including protection against power disruption and against viruses, Trojan horses, worms, and other methods of disrupting systems. In the chapter discussing goals, one of the points noted was that of asset management: making sure that the data, information, and knowledge that are produced are protected from loss and put and put in a readily used format. Security is a partial response to that goal.

One of the most basic security measures is **access control:** who can get to what data and under what circumstances? Among the minimum requirements are

Requirement	Pros	Cons
Password protection	Used properly, provides first level security and identification of users	Slows access to system, people may object to one more item that gets in the way of work, easily overcome, people forget passwords
Deny remote access	Limits access to those physically inside the lab, removes one potential source of data theft and alteration	Limits people's ability to work, removes the possibility of people traveling or working off-site from access to system
Call-back on dial-up access	Permits remote dial-up access, with logging of locations called, provides a means of control on remote access	Delays access for a short period of time
Key-card controls	Limits access to secured areas	Additional hardware/software required, can be overcome, requires monitoring and enforcement, not practical within a lab

Access controls are the least popular security measures and the first type expected in any system. The goal is to control access to systems and reduce the likelihood of theft and corruption. In order for it to be effective it has to be enforced. Whoever has that responsibility is going to be unpopular. They will be viewed as enforcing administrative measures that "get in the way of real work". Access control is necessary, should be enforced, and people need to be informed/trained in the need for and use of access control procedures.

Passwords are the most common form of access control. The first question is: do your systems have password control? Multi-user systems usually do, but how about the lab PCs? Are there routine procedures for changing passwords and choosing which are allowed and which are not? An editorial in the July 12th, 1993 issue of *PC Week* (p. 60, "Looking Forward") points out that an old situation is still current: people taking shortcuts to handling passwords by using the same password on multiple systems for example. Once one machine is compromised, all of its users' accounts are accessible. Avoid using simple words or names as passwords. One individual I knew used a deliberate misspelling of a word as a password—that was intended to thwart password cracking systems based on spelling dictionaries.

Passwords are a problem to the legitimate systems user because they deny immediate access to a needed system. The process of logging on and off each time you need access to a system is perceived as "wasting time". People will often not log off when they leave their desk for a few minutes to avoid the inconvenience of logging on when they get back. The problem is more severe when several people need access to the same computer and it has a provision for only a single password. The password at that point becomes commonly known and is useless. On small systems (PCs) there may not be options other than training and due diligence. On larger time-sharing and client/server networks more robust user-authentication systems may be supported.

The *call-back* is a useful means of providing and controlling remote access. The benefit is that users can gain access to the system from home or while traveling and provide control over who has access, keeping a log of where calls were placed from and a control list of which telephone numbers are permitted to be called. The process is simple: a user places a call to the system and provides identification. The system then logs the user off and calls that user's assigned number. The result is controlled access, a record of access, and users being provided access to the system. The system may be compromised by call forwarding, but that can be minimized by canceling the call after the second ring. The initial log-in process is also a point of vulnerability. It needs to be thoroughly tested so that people can't get past it—this includes eliminating any special access provided by system maintenance personnel.

Computer viruses, Trojan horses, and worms are pieces of software that can "infect" and cause damage to your system. A *virus* is a program that attaches itself to another program or data file and, when activated, can cause havoc in a system. Not all viruses are destructive, some just put messages on the screen, but all should be treated as undesirable factors. Viruses will replicate, spread themselves to other disks and software, and move across networks. Their effects can include crashing a system and erasing all data from disk. Once activated, a virus can set itself to be dormant and then awoken at a specific time or in response to an event.

A *worm* is a program that can carry our its activities independent of other software—it doesn't require a host. The much-publicized Internet problem of 1988, which affected thousands of computers, was caused by a worm spreading itself through a network.

Trojan horses are software programs that on the surface look like useful utilities or programs, but while they are doing their "useful" activity, they cause damage.

Viruses are the most problem, and incidences of their occurrence continue to increase. In 1990, 26% of the computer installations in the U.S. reported problems. In 1991, the number jumped to 61% (source: Dataquest, National Computer Security Association). Today it is rare to find a company that hasn't had an experience with an attempted or successful infection. The problems caused include loss of productivity (lost time due to rebuilding systems, or recovering data files), screen messages, corrupted data files, lost data, unreliable applications (intermittent termination), and system crashes.

The sources of these programs used to be limited to noncommercial software sources (downloads from bulletin boards, shared software, etc.) but now include infected commercial software. *PC World* recently (July 1993) received three demonstration disks from Microsoft containing the Forms virus; antivirus software detected the problems before any damage was done. This incident was originally reported in the July 19th, 1993 issue of *InfoWorld* (column by Robert Cringely, p. 98) and confirmed through private communications.

Data files shared over networks are another source of the problem. As networked systems become more commonplace, so will the problem. It is also possible to send viruses by electronic mail.

The following is a very brief list of some viruses and their effects (there are hundreds of viruses):

- Ambulance Car Virus—Ambulance randomly moves across bottom of screen
- Friday 13th—Erases programs when you run them
- Disk Killer—Causes unexpected formatting of hard disk, all data lost
- Stoned—Damages directory and FAT, results in data loss
- Datacrime—Adds garbage to file and reformats disk, data loss
- AirCop—Attacks boot sector on non–write-protected diskettes, system needs to be rebuilt, possible data loss
- Ping-Pong—Wipes characters off screen
- Falling Letters—Characters fall to bottom of screen

There are commercial and public-domain software packages that provide protection against these problems. Their use should be part of your system security SOP. All incoming software should be tested on a machine that is not connected to a network; if a virus is present, this step will limit its spread. Procedures like this, just as with any security procedure, are likely to be regarded as just another administrative requirement. Something to be endured. People need to be informed that these are serious matters and that it is their data, and their systems, that are at risk.

Virus protection packages are a good defense. This software needs to be kept up-to-date; the packages only protect against problems they know about and new bugs are constantly creeping up. Some commercial vendors have an update service that is worth considering.

There has been some debate about whether the virus scare, particularly when the news media becomes caught up in it as it did with the Michelangelo virus, is real or just a way of selling software. This is a real problem; the question is really one of magnitude and how likely you are to be affected by it. The measures necessary to prevent the possibility of becoming a victim are simple and inexpensive enough (some very good anti-virus software is free!) and the best approach is to be conservative, take and enforce precautions, rather than risk the loss of valuable information, time and effort.

SYSTEM BACKUP AND RECOVERY PROCEDURES Your building/laboratory just had a major disaster, everything was lost. Would your organization be able to recover? One of your laboratory data system's hard disks is damaged—how much data will be lost when it is replaced? What effect will it have on the lab's operations, its ability to meet the work schedule, or regulatory requirements? The answer to each of these points is regular and frequent backups (making copies of all data and programs) on to removable media. This should include periodic testing of the backup process to make sure that the copied data can be recovered and sending copies to off-site storage.

Backup procedures do need to be tested, including regeneration of a system from the backup media; how else will you know if it really works? This needs to be done periodically so that changes to the system are reflected in the backup procedures. If not, you may think you have a good backup, but be unpleasantly surprised when the backup is needed. Backup should be done to two different types of media (floppy, removable Winchester, DAT tape, etc.). If one medium on a computer is faulty, the ability to recover will not be hampered.

Backup is among the least popular system maintenance activities, and one of the first to be overlooked in routine system work. Schedule pressures, overwork, "I'll do it tomorrow—everything should be all right for one day", these all cause it to be overlooked. If the system is on a network, then have backup over a network be part of a daily routine—eliminating it from the workload of each individual. If that isn't possible, then drill the need for backup into everyone. Make it easy. Provide a tape or large capacity removable media so that the backup can be done in a single operation that doesn't require changing disks. The additional hardware is expensive, but so is lost data.

SOFTWARE AND SYSTEMS DEVELOPMENT Earlier in this chapter we looked at the points that need to be considered for the make or buy decision. Should you decide to build your own system, you should have a set of SOPs that describe the process of how it will be designed, implemented, tested, documented, etc. There are "industry standard" practices for carrying out a development project. Those listed below are those from the American Society for Testing Materials (ASTM). Others are available from IEEE Computer Society, the Association for Computing Machinery, and other sources.

ASTM Ref.	Title
E 622-84	Computerized Systems
E 623-89	Developing Functional Requirements for Computerized Laboratory Systems
E 624-89	Implementation Designs for Computerized Systems, Developing
E 625-87	Training Users of Computerized Systems
E 626-83	Evaluating Computerized Systems
E 627-88	Documenting Computerized Systems
E 730-85	Developing Functional Designs for Computerized Systems
E 731-90	Procurement of Commercially Available Computerized Systems
E 792-87	Computer Automation in the Clinical Laboratory
E 919-83	Software Documentation for a Computerized System
E 1029-84	Documentation of Clinical Laboratory Computer Systems
E 1113-86	Project Definition of Computerized Systems
E 1206-87	Computerization of Existing Equipment
E 1283-89	Procurement of Computer-Integrated Manufacturing Systems
E 1308-90	Identification of Polymers in Computerized Material Property Databases
E 1309-90	Identification of Composite Materials in Computerized Material Property Databases
E 1313-90	Development of Standard Data Records for Computerization of Material Property Data
E 1314-89	Structuring Terminological Records Relating to Computerized Test Reporting and Materials Designation Formats
E 1314-90	Rapid Prototyping of Computerized Systems
E 1338-90	Identification of Metals and Alloys in Computerized Material Property Databases
E 1339-90	Identification of Aluminum Alloys and Parts in Computerized Material Property Databases

Of those listed above, numbers 622, 623, 624, and 919 are particularly useful. You may choose to use practices described by another group, but that isn't an issue. The key points are that you have standards for systems development and that those standards reflect the current state of the art. The

OF PARTICULAR INTEREST

developers will probably howl at the amount of documentation required. That documentation is necessary if you are going to produce something that is going to be useful, be around for a long period of time, and be supportable. If your facility is covered by a regulatory agency, you really don't have a choice, assuming you want to stay in business.

While these procedures are a lot of work (and they are), they shouldn't be taken lightly. The process can detect flaws in a system, missing components, and inconsistency before the actual development process begins, saving time, money, frustration, and yielding a better product.

SYSTEM MAINTENANCE Periodic checks of computer hardware and networks are necessary. Preventive maintenance routines can test the computer system and detect problems in storage systems and other components; many can make corrections. Logs should be kept of the type and frequency of problems since they may point to a degrading system that may crash at an inopportune time causing data loss.

Hard disk drives need to be periodically defragmented to maintain high performance levels. Fragmentation—having data scattered over the surface of a disk, rather than having it laid out sequentially with one segment immediately following the next—occurs when files are created and deleted frequently. Disks begin with a low degree of fragmentation when first put into use and become more fragmented as files are updated and removed.

Figure 4-14 shows the data layout on a disk with no fragmentation, low level fragmentation, an finally a moderate level. In this figure the portions of the disk containing data are in black, and the empty areas are in white. The figure can be interpreted by considering a disk as being a long line of segments, like a line of boxcars starting at the top left corner, that has been folded when it hits the boundary of the illustration. The first line is filled, then the second, and so on.

Fragmentation is important because high levels (over 8%) will seriously affect system performance, slowing down disk data access. If real-time data acquisition is going on, data could be lost due to data overruns (not being able to finish writing data to the disk before new data becomes available). There are software packages that will clean up fragmented systems, and they should be used in combination with system backup routines. Backups should be done first, in case an error occurs during the defragmentation process.

Maintenance also includes software. Making sure the current versions are in use and that out-of-date versions are removed and archived (you may need them later if questions arise about effect of new features versus old software and whether or not something changed). Maintenance also includes the controlled introduction of software in the laboratory, this was covered in part in the earlier section *Planning for the introduction of revised/update software*, but is worth reviewing.

Commercial software developers build their products in a controlled environment. Assumptions are made about the operating environment, and

DEVELOPING AN IMPLEMENTATION PLAN

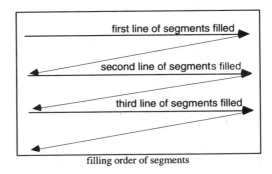

filling order of segments

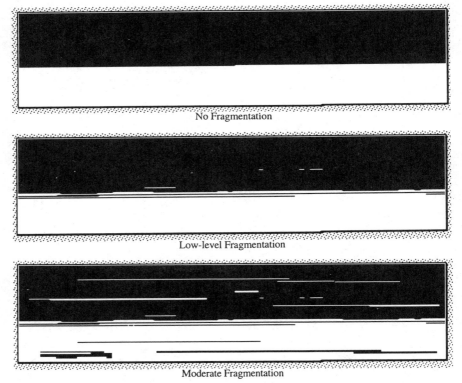

Figure 4-14 Data layouts.

those assumptions are frequently at odds with reality. Interactions with other programs may not be viewed as an issue, but they do happen. Screen savers can affect data acquisition. One individual reported that he was having problems with an accounting package, the numbers didn't add up right. As it turned out, a utility that altered the screen driver, making the displayed image

OF PARTICULAR INTEREST   **101**

larger, altered the math library and caused the error. Any piece of software, regardless of its source, should be tested before putting it into use. Not just tested to make sure critical functions work, but also make sure it doesn't interfere with the operation of other software packages. This is an expensive and time-consuming process, but basing decisions on faulty data is even more of a problem.

Standard operating procedures are needed for both the maintenance and introduction of computer hardware, software, components and systems.

SYSTEM VALIDATION   Because of the emphasis on regulatory issues, validation has become one of the more interesting topics in laboratory systems design and implementation. Validation is a response to a challenge: How do you know it works? Where "it" can be a procedure, a program, or the entire laboratory's operations. The validation process begins as soon as you decide to create a laboratory automation project, whether it be the entire lab or a component.

A validation program is both a cooperative and an adversarial process. It's cooperative in the sense that laboratory workers and system designers need to be responsive and open to comments, criticisms, and requirements for more testing or proof that something does indeed work. Those from within a company who are evaluating the effectiveness of a validation program need to understand that they are working with people, that they need to support their comments with data and facts (not just feelings or arbitrary judgements), and be open to the possibility that their findings may in fact be challenged and respond fairly. It is adversarial since a program is being evaluated first by an internal audit and then by outside auditors (who also test the internal testers—no one is safe!). The attitude of the validation/audit team is critical to the success of the program. If a we/they approach is taken, you are in deep trouble. The auditors/testers need to be a separate group, with a separate reporting structure, but they have to approach their work with a goal of helping develop a successful system, and not by keeping score of the number of problems that they have to fix. The two reporting structures need to join so that decisions can be made and put into effect if differences cannot be resolved otherwise.

The validation program needs to be designed in parallel with, and in cooperation with, the setting of goals and development of an implementation plan, several references on validation are noted at the end of this section (the list is taken from an article by Ken Chapman, in the *LA Forum*, Winter 1993 issue). Testing and evaluation needs to be viewed as part of the design and implementation and not as an afterthought. This will reduce the cost of the program and improve its effectiveness by testing the design as well as the final result and by catching problems early enough to be corrected without sending the entire facility and staff into chaos. Ideally, the validation team should be made up of senior developers or managed by senior developers who have been through the process themselves. If that is not possible, then time should be spent on team building and generating cooperative attitudes–Technical problems can be solved by work. Interpersonal problems that become political issues can kill projects. While this may be easier done in a large organization with a budget to handle additional personnel, smaller companies need to have a similar approach. The validation role may be supplied by an outside firm. One person cannot do the design, implementation, testing, and validation.

The answer to the question of "how do you know it works?" is a summation of the design, implementation, product evaluation/selection/testing, and maintenance SOPs. Properly laid out, they should provide an answer at each step of the process. Let's consider the purchase of a hardware/software package to carry out an analysis.

- The first step is to define why a product is being purchased and what is it intended to do.
- Then: How does it relate to other systems in the laboratory?
- How are linkages—if any—made to other systems? How are the linkages tested? How will you know that the existing, validated systems haven't been adversely affected?
- How is the choice of products to be evaluated made?
- What is the basis of the evaluation, what are the product selection criteria, and what trade-offs are permitted?
- How has the product been tested, and what is its history of use (new products to the market are going to require significant testing on the part of the user)?
- Who is responsible for the installation, and when it is installed how will you know that it is functioning properly?
- What periodic checks are required to ensure that it will continue to function?
- Are SOPs in place for its use, maintenance, and people's training?
- Has it been tested in conjunction with other related systems?
- What are the final acceptance criteria?
- Under what conditions does it get put into productive use?

Many of these questions—and this list may be incomplete—have been dealt with unconsciously in the past. You didn't put them down as a formal process; you did them out of habit or common sense. The validation process requires that that unconscious process be formalized. That the decisions and the rationale for them have to be documented and reviewed. That review, which is not necessarily intended to challenge the correctness of the choices, but the completeness and accuracy of the documentation is a needed activity—you might not be available when questions are asked. An "unforgetting" documentation mechanism is needed to record decisions.

The end result of all this is a ready answer to the initial challenge.

REFERENCES

FDA, 21 Code of Federal Regulations Part 211, "Current Good Manufacturing Practices for Finished Pharmaceuticals", *Federal Register* 41, 316878–6894 (1976).

K. G. Chapman, "A History of Validation in the United States, Part I", *Pharm. Technol.* 15 (10), 82–96 (1991), and "Part II, Validation of Computer Related Systems," *Pharm. Technol. 15* (11), 54–70 (1991).

FDA, "Guide to Inspection of Computerized Systems Used in the Manufacture of Drug Products" (the *BLUEBOOK*) Feb. 1983.

ANSI/IEEE Standard 729—1983 Glossary of Software Engineering Terminology, 1983. Standard 730.1—1989 Software Quality Assurance Plans, 1989. Standard 828—1983 Software Configuration Management Plans, 1983.

ANSI/IEEE Standard 828—1983 Software Test Documentation, 1983. Standard 830—1984 Guide to Software Requirements Specifications, 1984.

International Organization for Standardization, Geneva 20 Switzerland, International Standard ISO 9000 3, *Quality Management, Quality Assurance* Standards—Part 3: "Guidelines for the application of ISO 9001 to the development, supply and maintenance of software"—First Edition 1991–060 1.

National Center for Drugs and Biologics and National Center for Devices and Radiological Health, Guideline on General Principles of Process Validation, Rockville, MD, May 15, 1987.

PMA Computer System Validation Committee, "Validation Concepts for Computer Systems Used in the Manufacture of Drug Products", Pharmaceutical Technology, Vol. 10, No. 5, May 1986, pp. 24–34.

PART 2

TECHNOLOGIES FOR LABORATORY AUTOMATION AND COMPUTING

CHAPTER

5

Technologies for Implementing the Lab Automation Model

The earlier chapters of this book have described a model for laboratory automation, the need for a strategic approach and the issues in developing an implementation plan. Those are essentially paper-and-pencil exercises that deal with planning and goals. Eventually they have to turn into something real that actually works. The point of this material is to review the technologies that are available for laboratory automation and computing, how they have been applied, and what can be done to improve their usage.

We will be using the laboratory automation model (Chapter 2) and the knowledge/information/data model (Chapter 3) as a means of organizing and coordinating the material. When you are reading this material, there are some items to keep in mind. First, not all of the technology described will be of use to you. Second, the instruments, devices, and computing hardware and software are intended to be aids in managing your laboratories and not vocations in and of themselves. Finally, each of the technology elements has to be considered as part of a system that has to be validated; it is necessary to keep track of the linkages to other equipment and the flow of data to and from the devices. Much of what is described will have a quality control/testing/analytical laboratory flavor. That is a combination of my background and the fact that those types of laboratories are the most complex from the standpoint of applying automation and producing an integrated system. Almost everything that is described should be applicable to research facilities.

This material will be covered in two dimensions. First we move across the LASF model from left to right (sample storage management to laboratory management systems). Then having completed that pass, we look in more detail at the technologies that laboratory computing packages are built upon—the layers that make up the knowledge/information/data model.

As we progress on the first pass we will find that we are looking less and less at concrete concepts and technical issues and more at people and organizational issues. In principle, issues such as robotics, instrument interfacing, programming against well-defined criteria are engineering issues, and we believe (hope?) they can be resolved with study and work, dealing with a more objective world of physics, chemistry, and logic. On the right side of the laboratory model (working primarily with information and data) we run up against "soft" issues—people and the organizations they work in. Computer systems by themselves in well-understood applications are still interesting problems. Applying computing technology to organizational behavior problems is rather challenging. As a result the comments as we progress will become less certain and defined and rely more on 'you really need to think this through yourself' more and more. That latter point is not an attempt to avoid the issues and sweep the details under some rug. Doing data acquisition from an experiment (measuring blood pressure in a laboratory animal or recording the movement of a pendulum) is a place where we can discuss the trade-offs of different approaches, but the problem is essentially the same in one lab as another. Organizations are different, and legal aspects aside, they are made up of people and are dynamic. Issues, structure, and priorities change; sometimes very quickly if a new manager comes into the picture. One group in one part of a company can be managed differently than another. Corporate cultures vary. Those are the unknown variables that have to be understood, and the items that will have a significant (perhaps the most significant) bearing on system design and success. System design and implementation is a matter of group psychology as is hardware and software engineering. Ignore that point at your own peril.

SAMPLE STORAGE MANAGEMENT

Sample storage management is the first stop in a laboratory and the most recent in drawing the attention of vendors. This isn't surprising since an automated facility is going to be a custom designed system and not a mail order item. The sample (solid, liquid, gas, mouse, etc., numbering in the tens to thousands) has to be put somewhere, have its environment controlled, be able to be retrieved and stored, and maintain a chain of custody. There are other factors as well, but that will do for a start.

The typical nonautomated storage facility can be likened to a small grocery store or market. Perishable or temperature sensitive materials are put in cooler/refrigerator/freezer, and more stable materials are placed on shelves. There is usually some underlying order to material placement, but exceptions exist. A lot of time can be lost finding material, re-organizing shelves periodically, removing out-of-date and completed samples, and so on. All this is human effort and expensive. That problem is magnified considerably in clinical studies and material stability evaluations.

Since the facility described above is manually managed and controlled, all transactions are manually entered in notebooks or laboratory management

SAMPLE STORAGE MANAGEMENT

systems. Each step in a regulated environment has to be checked and validated. An alternative to this method is offered by Automated Systems Integration Corporation (Camp Hill, PA). It and other vendors as well offer an automated sample handling system that relieves most of the above issues. Figure 5.1 shows a top-view and perspective drawing of an existing system.

In this particular instance, the facility contains 3 sets of 24 revolving shelf units, each of which is 8 feet high. Each shelf unit holds 9 sample trays. The entire facility can be maintained at a specified temperature and humidity.

When a particular sample(s) is required,

- The sample information is entered into a workstation which directs the collecting mechanism
- The trays containing the required samples are pulled from the carousels and delivered to the operator using the conveyor at the bottom of the top view (at the left of the perspective view).
- A barcode scanner identifies the tray, and a map of the tray is displayed on the workstation screen. The map shows the location of each requested sample. Each sample is barcoded (in addition to the tray), and when the sample is removed, it is scanned and the fact that it has been taken is recorded—the slot that contained the sample is marked as empty.
- Returning a sample is just the reverse: scan the sample to identify it to the system, identify the slot and tray that will hold the sample, and return it to the storage unit.
- Records are maintained for each sample including who requested the material, with data and time stamp.

When samples are completed and need to be disposed of, a list of materials for removal is given to the system, the trays are sent to the operator, material removed, and their locations in the shelving units are marked as empty.

Overall, the system provides better use of space, control over material, and a chain of custody. It also improves safety for personnel since they won't be carrying material and climbing over shelves. The same type of structure could be used for some biological materials, though live animals may object to movement (it wouldn't be difficult to design an arm moving on a pair of rails or one on the ceiling and one on the floor to pick-up and transport individual cages). With a considerable amount of software and hardware development, you could consider extending the conveyor system to one that delivers samples to instruments lined up along the belt. Systems like that are common in manufacturing plants and large-scale quality control labs can be considered as sophisticated production facilities whose output is data and information. That viewpoint could have interesting ramifications if applied to a new laboratory; trade-offs would have to be made between efficiency and

TECHNOLOGIES FOR THE LAB AUTOMATION MODEL

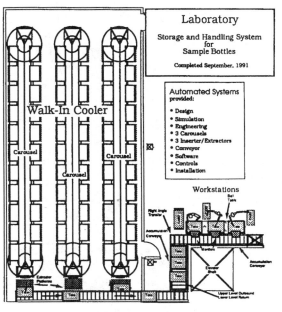

Top View

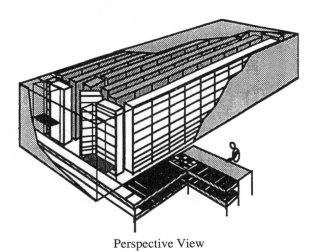

Perspective View

Figure 5-1 Automated sample handling system.

flexibility. Laboratory automation on a large scale might be attractive for contract labs that do large amounts of limited types of tests.

The benefits of this type of system increase when the list of samples is provided by a laboratory management system (LMS). The LMS would send a list of samples for a particular study or test, have the technician or operator pick them up (or remove need portions), and then send the rest back into storage. The system is designed with simulators to remove bottlenecks and improve performance before it is installed.

The implementation issue that needs to be taken into account is the linkage between the LMS (which could be part of a LIMS or a sophisticated commercial multi-instrument data station). File formats, method of transmission, and message structure are key considerations.

Extending this line of thought into large stability laboratories would be interesting. Particularly if samples were placed in and removed from environmental chambers automatically. The robotics systems required would not be too different from those used in large-scale tape storage facilities. Storage Technologies has a tape librarian that places tapes on the inside of a hollow cylinder and uses a robot that can read barcodes to select cartridges. Although the initial outlay for designing and building such a system might be expensive, automated test stands for carrying out tests could provide more data and earlier failure detection than manually operated systems.

SAMPLE PREPARATION—LABORATORY ROBOTICS

Laboratory robotics developed out of a need to reduce the amount of manual effort expended in preparing sample for testing. Since it came to the market (over ten years ago), the products have divided into two camps: dedicated systems (usually integrated with an instrument) and general purpose devices that require design and programming on the part of the user.

Dedicated devices include automatic samplers and auto-injectors and machines such as Zymarks Benchmate series. The Benchmates, though very flexible, are designed to perform a limited range of fixed function operations. At the other end of the spectrum are general purpose robots such as those provided by Zymark (Zymate series) and Hewlett-Packard (ORCA product line)—see Figures 5-2 and 5-3. The distance between the two ends of the spectrum is as wide at the comparison between a turn-key accounting program and a general purpose programming language on a computer. In one case you get something ready to go, right out of the box, with a fixed range of functions. In the other, you get the ability to do any calculations you want: all you have to do is figure out how to put the pieces together. Vendors have recognized the need to simplify some functions on the general purpose systems and are supplying modules of hardware and software to carry out particular functions, such as screwing the cap on the bottle, similar to providing a subrouting library for a programming language.

The general view of laboratory robots is that they were a replacement for manual actions: mixing reagents, filling tubes, weighing samples, and inserting test cells into instruments. As a result their role in an information strategy has

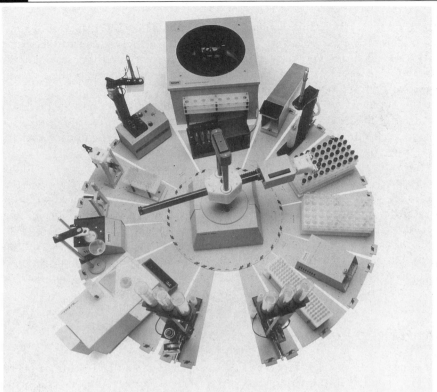

Figure 5-2 Zymate general purpose robotics system. (Reproduced by permission of Zymark, Inc.)

all but been ignored until the past two years—though, as always, there are exceptions... but we're jumping ahead. The typical justification for a lab robot, even today, is based on person-hours saved doing repetitive tasks and faster sample throughput. An article by Frank Zenie, a good one for justifying a robotics system, presented at the 1992 International Symposium on Laboratory Automation and robotics (ISLAR, pp. 27–46 of the proceedings) is based primarily on labor savings. The appendix of the paper is a series of case studies illustrating increased sample throughput and cost savings from 11 companies. In one, the number of assays per person increased from 445 to 921 over a three-year period, another offered cost savings of $411,300 in contract lab fees over a two-year period. Others describe a payback period of four to six months as a result of cost savings. Properly implemented and validated, laboratory robotics can produce significant cost savings for well-designed and well specified applications. They can also become a major financial sinkhole for poorly thought through applications—just like computers. In addition to labor savings, the justifications should, and do, include the human element: people are freed to do more interesting and challenging work.

SAMPLE PREPARATION—LABORATORY ROBOTICS

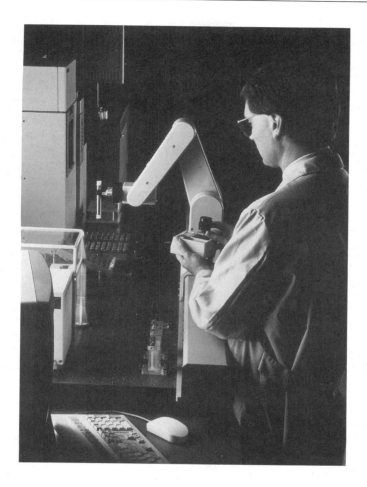

Figure 5-3 Technician teaching ORCA. (© 1993 Hewlett Packard Company; reproduced with permission.)

Robotics importance doesn't stop there. In fact, its major contribution to laboratory work has yet to be realized, as does its role in an information strategy.

The typical output of a lab robot is a printed report or a disk file that has to be processed by another program; a situation that's similar to most data stations. In terms of the cost/value relationship of data, lab robots are no different than any other laboratory data station. We can do much better than that. In order to increase the gains from an investment in this technology we have to change our perspective.

From the vendors' viewpoint the robot is the center of its universe. Everything in and about a robotic workstation is there with the assumption that the robot's programming is in control of the pumps, balances, instruments, and other equipment that fit in its reach. Suppose the robot's controller were subservient to another device with more capability and access to more information? That question is getting more attention. A few papers at the 1992 ISLAR conference began to address that point as did an earlier article that I will cover in more detail (J. Brosemer and J. Liscouski, "Computers and robotics: a synergistic system", *American Laboratory*, Vol. 18, no. 9, Sept. 1986, pp. 80–83).

The point of this article was to show how the use of a computer mated to a robot could be used to extend the latter's capability. It was not so much an example of a real-world experiment as it was a demonstration piece. The basic experiment was a titration, something that the robot was capable of doing by itself, except that the end result of the device's effort would have been the volume of titrant at the end point rather than the full titration curve that the computer generated. In addition the stand-alone robotics system would have added titrant in fixed quantities, whereas the computer-assisted device calculated the amount of titrant for the next addition based on the slope of the curve. Communications between the two units was through a serial communications port. The nature of the communications was the key element: we used a Zymark Z845 computer interface that allowed the external computer (a Digital Equipment PRO380, a PDP-11 class machine) to act as a substitute for the keyboard and display on the Zymate system. This was fundamental to changing the control model for the combined systems.

As noted above, vendors have an egocentric view of their products, the robot and its programming are normally the controlling element. Programming a robotic system is done from the point of view of the robot as master and having it contain the master program for everything that occurs in its operational space. What we did was to load the robot with functional primitives—commands that the robot would execute on demand (add titrant, get sample, etc.)—but with no overall control program; it was a slave to the PRO-380. The PRO-380 contained a program that, based on the current set of conditions (slope of the titration curve as a result of the last addition), determined what the next instruction should be (add x ml of titrant, terminate test, make coffee, etc.). At each step in the process, the PRO-380 would determine what to do, send the primitive to the robot, which would execute a series of programmed commands to accomplish its short term task, and then, upon receipt of a "successful completion" response, go on to the next step.

From the standpoint of a simple experiment this had a few advantages over a stand-alone system:

- More information was available to the analyst—the complete titration curve could be examined, rather than just getting a single number.
- The programming could be much more extensive since the PRO380 had much more computing capability and data storage than the robot.

SAMPLE PREPARATION—LABORATORY ROBOTICS

- If needed, the computer could send a message to an operator (out of titrant, alarm conditions, etc.) over a network via EMAIL.
- Samples could be retested based on test results (out of specifications for example).
- Zymate primitives could be down-loaded as needed altering the functionality of the robot (this could cause problems in system validation since the operating capabilities are dynamic rather than static).

It was also a bit more expensive.

This extension of control to a higher-level device (one that has access to other databases, programming, messaging, interfaces to other instruments or robots, etc.) provides for much more sophisticated experiments and control structures; this is where the information strategy comes into play. How do you want to extend the robot's capability, changing it from a reporter of point data elements to an interactive generator of certified results?

Given this higher order of control, the data presented to the analyst/researcher can have a higher value. Out-of-specification result can be rerun based on data from LIMS. If the results are duplicated, then one or more standards can be run and verified to show that the sample really is out of spec and that the system is performing properly; the increased value of the data comes from this added confirmation, which would cost little to obtain and be more timely if done automatically rather than waiting until it was spotted in a report. If the result didn't duplicate, an additional set could be run, and if problems persisted, corrective action could be taken including

- checking standards and calibrations to determine if the system was operating within established parameters
- stopping testing and notifying someone that the system is not operating properly

Not only would we have identified a problem, but reasonable diagnostic/corrective action would be taken, which would be particularly appreciated on the midnight to 8 am shift.

In addition to checking for out-of-spec material, statistical quality control procedures could be programmed in to check on all samples, giving a higher level of reliability. Although this is going to sound like typical sales hype, the result could be logged directly into a LIMS or queued for entry, pending review and approval; all it needs is "a little programming."

The control system could coordinate the robot's sample preparation with a scanning spectrophotometer or other device too complicated for the robot's control systems to cope with. In short, rather than considering a laboratory robot as an isolated workstation, it should be examined in terms of communications and data management requirements and capabilities and not as a stand-alone device. This point is expanded on in an ISLAR article by Stephen

Metzner ("Data vs. Information in the Automated Laboratory", 1992 ISLAR Proceedings, pp. 88–97).

The key is communications between the robot controller and the higher-level computer, and that is where the problem lies. Instrument vendors rely on serial interfaces for communications, frequently with no or limited protocols for error detection and correction. In order for systems like the one described above to function, both the robot and the control system need to be able to ensure that messages are sent and received properly, and that errors can be detected and corrective action taken. Without that guarantee, the entire control structure is at risk and subject to failure. It is also impossible to validate. If you are interested in pursuing systems like this, the communications structure and capabilities have to be examined in detail.

Ideally, the communications protocol should have a means of positively determining the validity of a message and taking corrective action on errors. Checksums or other mechanisms are useful. The NIST/CAALS organization is releasing a draft of its communications specification for instruments and that should be useful in these applications. Much depends upon the nature of the robot controller. A general purpose computer may have network access and control or interapplication communications (Dynamic Data Exchange on MS-DOS/Windows, or the Macintosh communications system). Proprietary controllers will have whatever mechanism the vendor provides.

In the past, some instrument manufacturers have relied on the validity of the content of the message as a means of determining communication errors, thus avoiding the need for a formal protocol. For example if a number is expected, the string of characters representing that number is converted to a numeric value. The conversion routine is used to check the validity of the data. That isn't reliable. There is a non-zero probability (depending on the length of the string) that a one-bit error can occur and not be detected. Not only will you get bad data, you won't know about it.

Robotics Technologies

There are two different approaches available from general purpose robotics vendors. One is typified by Zymark in its Zymate series, the other by Hewlett-Packard in the ORCA product line. The most obvious difference is the working range of the devices.

The Zymate systems have been on the market the longest of the general purpose lab systems. The systems have developed steadily in response to customer needs. The robotic arm sits in the center of the work area (the controller is a separate module), which describes a cylinder. Equipment is arranged around the arm base in a circle (see Figure 5.4). That equipment can consist of balances, centrifuges, instruments, solvent delivery systems, injector stations, mounts for replacement hands (the design of each varies by use), tube racks, etc., all depending upon the experiment in progress. In addition to moving the arm, the controller has access to switch modules that control other

SAMPLE PREPARATION—LABORATORY ROBOTICS

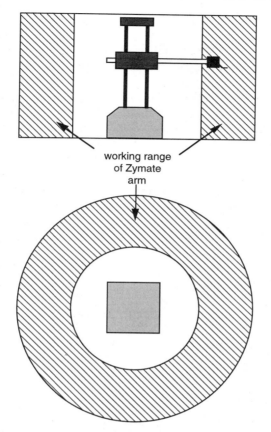

Figure 5-4 Zymate system.

devices either in the immediate vicinity of the arm or somewhat removed from it. The center of the working area is empty to allow room for the arm to move.

This arrangement has proven to be effective for a wide range of applications. Communications between the controller and other devices is via a proprietary serial protocol. One of the interesting aspects of the interface scheme is that it is designed to remove interrupts from the system. No device can interrupt the controller, which greatly simplifies programming and increases reliability. If an external device generates an interrupt, that signal is trapped by the interface box, which takes whatever action is necessary and holds data until the controller polls it.

One issue that occurs in setting up a robot system is locating the equipment that the machine is supposed to access. It has to "know" where things are in a three-dimensional space so that it can get to them. In early systems, a fair amount of time had to be spent locating devices and specifying locations around the device as reference points. Through the development of Pyetechnology, that problem has been neatly solved. Each standard component (test tube racks, hand mounts, etc.) is attached to a pie-shaped wedge that attaches to the base of the robot. Software modules are included for addressing the devices. Locating them is a matter of identifying one or two reference points, and everything else is calculated from those.

Zymark has added new capability to its Zymate series of robots. While most current installations are based on the fixed-based unit, the company has extended the robot's reach by putting the base on a track—prior to this development, track-based Zymate's were implemented by the users. The length of the track can range from a few feet to over twenty in length. This gives the robot the ability to manage and use much more equipment, and more complex techniques—complete bioassay systems can be easily accommodated, with options for different types of assay procedures. The photograph (Figure 5-5) shows such a typical setup. The robot is just to the right of the center of the highlighted oval. The dark area immediately below the robot is the track.

Hewlett-Packard's approach is different than Zymark's. Their movable arm is always mounted on a rail that can be up to 2 meters in length (Figure 5-6).

While the ORCA is a general purpose machine is every sense, its origins with an instrument vendor are evident. The rail makes it easy to access instruments and their subsystems lined up on a bench. It does have an array of I/O functions including support for HP-IB and RS-232C (Figure 5-7). The HP-IB interface is particularly useful with HP instruments and accessories. The

Figure 5-5 Typical setup of Zymate system. (Reproduced by permission of Zymark, Inc.)

INSTRUMENTS AND MEASURING DEVICES

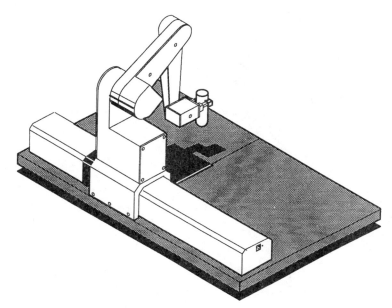

Figure 5-6 Hewlett Packard movable arm.

control software is based on MS-DOS and Windows. This gives considerable flexibility, but you need to be careful. Windows does support multiple applications, and as noted earlier it is possible for them to interfere with one another. Later versions of Windows *may* relieve this issue as pure 32-bit implementations of the operating system are designed. Other vendors have variations on these themes.

So far we've looked at two areas in the lab model, both of which are concerned with handling of materials. The next section is on instruments and evaluators and begins to bring us into the knowledge/information/data model at the entry point of the "data" database.

INSTRUMENTS AND MEASURING DEVICES

This is the point at which spectrometers, chromatographs, and experimental control systems reside. Until the past few years, the laboratory instrument industry has had a few common characteristics: each recognized that automation of the instrument was a key element of its business, most migrated to IBM PC compatible systems (the major exceptions were mass spec vendors that required UNIX workstations), most used control over data storage format as a means of maintaining customer loyalty (there were several who provided an export mechanism to spreadsheets or other file formats).

The direction today is quite different. Standards are being developed, and

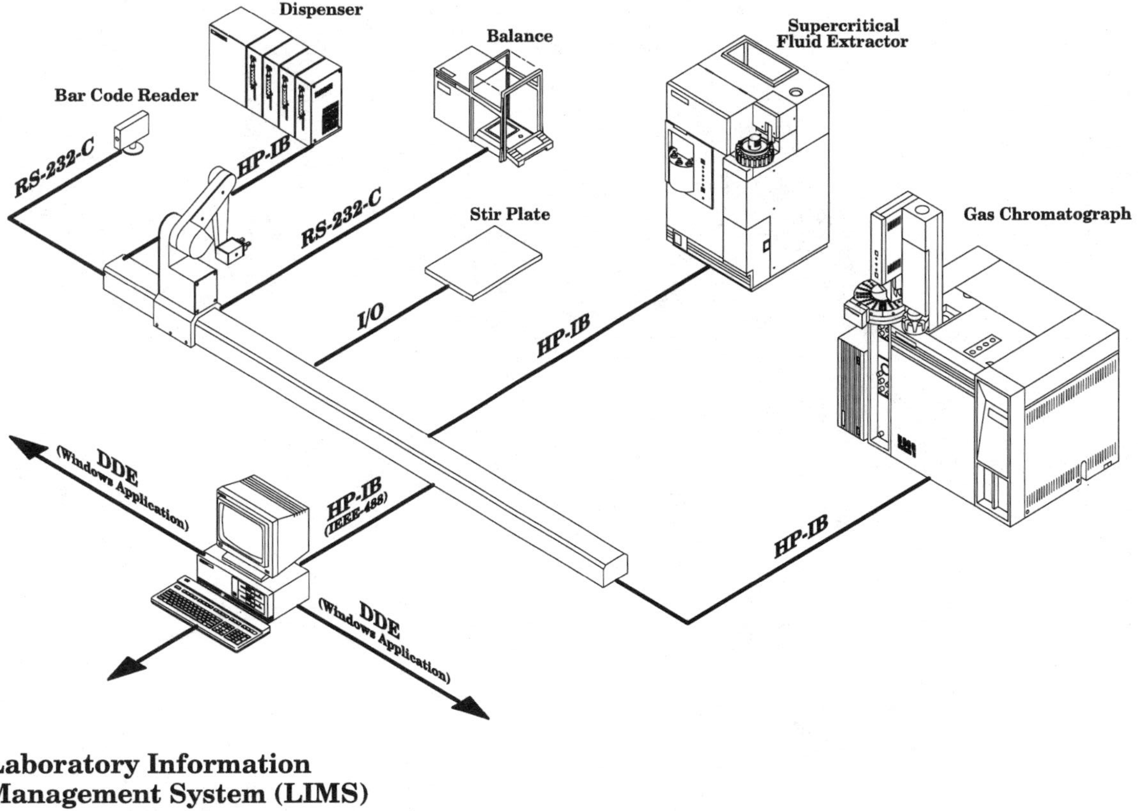

Figure 5-7 ORCA interconnections with laboratory instruments. (© 1993 Hewlett Packard Company; reproduced with permission.)

INSTRUMENTS AND MEASURING DEVICES

the marketplace is beginning to become aware of and understand their use. The Analytical Instrument Association has produced the Chromatography Data Format Standard, which has been in place for over a year. In the fall of 1993 the mass spectroscopy Standards is expected and an infrared spectroscopy Standard should be available in the fall of 1994. The implications of these standards' efforts was covered in Chapter 3.

Laboratory automation had its beginning with instrumentation. Everything we have today has evolved (not been designed!) from efforts to improve the scientist's productivity at the instrument and to meet the competitive pressures from other vendors who were successful at it. The next section is going to review some historical background on automation, which is of use if you want to understand how we got where we are and the barriers to moving ahead; otherwise you can skip to the next section, no one's watching.

Historical Digression

Everything started with the instrument. Up through the early 1970s, most vendors spent their R&D budget improving detector sensitivity and resolution—whatever was appropriate to make the device work better, do more, and be more attractive to the customer. Unless a new technology develops there are limits to how far you can push basic chemistry and physics in a commercial environment. The next major step to improving sales was to begin addressing how the instrument was being used, where bottlenecks occurred, and where user needs might be satisfied by new products.

At the same time researchers (in universities and corporations) were looking at computers—which at that point were becoming powerful enough and affordable—as a means of doing automated data acquisition from analog devices. The potential of using a computer for this work wasn't lost on computer manufacturers, and laboratory versions of systems began to appear, but the user was responsible for programming everything. There was very little commercial software available. There were mass spectroscopy systems on the PDP-12 and chromatography software on PDP-8s.

In some disciplines, automation asserted itself through automated sample injectors used to relieve the tedium of people moving samples through chromatographs. These auto-injectors, and their counterparts in other techniques such as the AutoAnalyzer from Technicon, made it easier to process samples through the instruments, but not process the data—that was still a hand and calculator/slide-rule effort. The combination of automated samplers, instruments, and data systems raised the possibility of fully automated instruments from sample to report, increased throughput, unattended operation (off hours!), and made lab managers salivate (sound familiar?).

Automatic data acquisition, analysis, and reporting was the wave of the future. The late 1970s and early 1980s saw a rapid evolution in equipment, the most aggressive was in chromatography. The advent of the microprocessor caused a fast turnover in product design and introduction. The directions in product development reflected the commercial orientation of the vendor. Instrument vendors tried a variety of tactics, including dedicated integrators that were functionally part of the instrument, integrators that would only work on a specific vendor's equipment, integrators that would work on anyone's instrument, and so on. For the most part, the basic premise was one instrument, one integrator. Vendors initially viewed the acquisition/analysis/reporting sequence as a means of automating calculations and getting through a lot of number-crunching and not as data/information management issue.

Computer vendors, wanting to sell computers, had a different perspective on the problem. In order to justify the cost of a computer for the same problem that the instrument vendors were attacking, they had to couple several users and a number of instruments to the same machine to get comparable cost-per-device numbers. Now we had an interesting competitive problem. Two different philosophies going after the same market. One stressed "we made your instrument, who better to automate it?", the other stressed the information management aspects and higher level functions ("it's a computer, you can do anything, just a matter of programming"). Alliances, such as that between Waters and Digital, looked like a combination of the best of both worlds. It would have been an interesting battle to watch except for one item, Intel. The Intel 4004, gave way to the 8008, the IBM PC was born, and the market gravitated to it. So the situation sat, at least for chromatography. Other instrument techniques followed similar paths.

At the same time, another bottleneck was being attacked—sample preparation. The auto-injector got the prepared material into the instrument, but preparation was still human-intensive labor. Robotics became part of the laboratory automation landscape.

Vendors began to realize that the data and information management issue was going to be a key element in the laboratory. Their goal was to make as much money as possible by selling equipment and services (cynical maybe, but realistic). One way to do that was to put a stranglehold on the customer by controlling access to data. Software vendors (including some instrument manufacturers, some of who sold computers too) saw the laboratory information management system as the means of obtaining the customer's undying loyalty. Others were less lofty ambitions, and budgets, sought to control the data from a single class of instrument (multi-instrument chromatography systems for example). Some did both, either by design or by acquisition. Computer vendors, not wanting to be left on the sidelines, took their own stab at the same issue. Some either developed LIMS of their own, or formed alliances with companies that had them running their machines. Their basic tactic was to ignore the application—the software house did that work—and try to own the underlying database product and the means of accessing it (own the network, and you own the communications; if you control that, you control everything).

The vendors did, and still do, provide good products that do a good job. Those products did, and still do, make laboratory computing easier and have

INSTRUMENTS AND MEASURING DEVICES

eliminated a lot of manual repetitive work. It all began with the expansion of the instrument business, with each vendor having a very strong sense of turf—and the need to maintain proprietary data structures to engender customer loyalty. Which brings us up to where we are now, and a changing landscape. Market dominance based on hardware—application or networking capability is short lived. The only clear industry leader today is Microsoft, and in that laboratory, the leadership stops at the operating system. Of more serious concern is being able to adjust to new platforms (both hardware and database) and maintain high levels of data access in a validated system.

The barriers to implementing an integrated laboratory automation system come from several points with a common theme: a desire for market dominance, and protection of an existing customer base. A fiercely competitive market is enough of a problem; add a serious recession on top of that and vendors are going to be concerned with survival. Vendors need to overcome their tendency to use control of data access as a means of preventing customers from changing their source of supply. The fact that the data format standards efforts were fostered by the vendors is an indication that that change is beginning. Consumers have a responsibility as well in this process of change. They—*You if you're a customer of lab system vendors*—need to tell vendors that you want standards developed, used, and supported, that you won't buy products that don't support or use industry standards. There is some resistance to the standards work. Some vendors don't believe that consumers care, or that standards are important to you. Some will suggest that they can do as good a job of system integration without standards as others claim to be pending their development—*just give them a large consulting contract for a custom developed solution.*

At this point instrument vendors have taken the initiative and begun to create data interchange standards; something the user community should be quite happy with and actively support.

The AIA's standards efforts are just one part of the work that is going on, which will be covered in more detail shortly. There is another, equally important effort underway at the National Institute for Standards and Technology (NIST) called CAALS (Consortium for Automated Analytical Laboratory Systems). Where the AIA standards work is directed toward the development of data interchange specifications between software packages, the NIST/CAALS work is designed to standardize communications between subsystems within an instrument workstation. Both are needed to fully develop the model we have used in this book. We'll take a look at both these efforts, beginning with the NIST/CAALS standards.

NIST/CAALS

If you are going to fully integrate automation in a laboratory, one area that needs attention is the instruments and apparatus themselves. All electronic laboratory equipment has certain characteristics: they can be turned on and off, they can be set to particular sets of operating conditions, and, they provide feedback to the user on their current state through lights, meters, alphanumeric displays, etc. Their ability to communicate electronically and be controlled by a computer varies widely. Some have no access (ovens for example), others are fully compatible with computer control. Full automation capability will require a change to that situation. And that change is desirable.

The benefits range from power savings (having a supervisory computer control equipment by turning on and off ovens, etc.) to full experimental method implementation (telling the computer you want to run a particular experiment and having it set up all the apparatus—*this would be very useful if a statistically designed set of experiments were being run!*). In the electronics industry, thanks to the development and *implementation* of the IEEE-488 bus and its follow-on standards, it's possible to set the operating characteristics of a frequency generator, have a high speed data recorder capture data, store it on disk, display it, plot it, open and close switches, all under computer control, all with a common interfacing method and protocol. Chemistry and biology labs, for the most part, can't. The major problem is communications.

Aside from vendors such as Hewlett-Packard, DuPont, and some Varian equipment, vendors have avoided the IEEE-488 bus and used serial communications (RS-232C). Some are implementing EIA-485 (*Note:* the EIA designation replaces the older RS abbreviation). Error detection and correction protocols are minimal, and so as a result is the capability of the communications channel.

The NIST/CAALS efforts, documented under CAALS-1 Communications Specification is similar to the ISO's Open Standards Interconnect model for computer communications. Not all layers of the model are required; some simple devices may only include the basic message protocol. The intent is to provide a standardized message structure that all laboratory devices will use, regardless of the communications link, to exchange commands and data. The result is that laboratory computers will have greater control over instruments. That it will be possible, when the standard is completed and implemented, to call up an experimental method that completely describes a study or analysis and have the computer set and verify instrument conditions.

This effort is still in the specification and development stage. If you would like more information, contact Dr. Gary Kramer, National Institute for Standards and Technology, Center for Analytical Chemistry, Bldg. 222 M/S A-343, Gathersburg, MD 20899-0001.

The AIA Data Format Standards

The problem that the AIA had to address was an interesting one: develop a means of exchanging data between vendors who are using a variety of hardware platforms, operating systems, languages, and numerical data formats. This looks like the situation described earlier where you have a number

INSTRUMENTS AND MEASURING DEVICES

of independent researchers working from a common database of data. There are a number of options for moving a file of data between machines and operating systems; the problem is making sense of the contents. That problem exists with word processors and spreadsheet software today, with the current solution being filters implemented by vendors and third parties (filters are small programs or software modules that convert data from one format to another with each filter specific for a particular pair of source and destination formats). The problem is the large number of modules that need to be supported and the need for consistency between each implementation. Every time Microsoft Word is updated, new filters are required or old ones need to be updated. For data exchange in laboratory environment (which must be validated—a significant problem when dealing with frequent revisions) this isn't acceptable.

The solution to the issue was a software package called netCDF (Network Common Data Form). NetCDF was developed by the University Corporation for Atmospheric Research and is supported by the Unidata Program Center in Bolder, COL. Funding for that support is provided by the National Science Foundation; the package is available in the public domain. Detailed information can be obtained from Unidata (request the netCDF Users Guide, Russell K. Rew, Unidata Program Center, UCAR, P.O. Box 3000, Boulder, Co. 80307-3000).

NetCDF is not a database, but a means of encapsulating and exchanging data between systems. It does not support the common database functions of sorting, indexed searchers, and so on. The goals for the development of that software are

- to provide a common interface for Unidata applications and data
- to provide an interface for both the C and Fortran programming languages (*Note:* the Fortran language implementation is through "jackets", a mechanism that converts Fortran subroutine calls into C language calls—this requires a common interface between the languages, something that is not guaranteed on PCs)
- provide random access to self-describing, network transparent, multidimensional data (netCDF files contain information about the information in the file, as a result it is referred to as self-describing; through subroutine calls you can ask for listing of variables, types of data, and data elements themselves)
- improve access to and use of scientific data
- increase the reusability of software

This was a good match to the requirements set up by the AIA, which was looking for

- independence for hardware platforms, operating systems, networks, and other proprietary constraints

- having broad applicability to standards other than chromatography
- not having significant support and software development overhead
- not being bound to a specific commercial vendor

There are capabilities within netCDF that make it useful for a variety of instrument data types and experimental situations (my chemistry background is going to show, apologies to nonchemists). Chromatography data is usually stored as a one-dimensional data set, a vector of elements whose temporal relationship to each other is described as a time increment. Spectrophotometry data is two-dimensional (transmission versus wavenumber). Those are relatively simple. The output of a diode-array detector for a chromatograph is three-dimensional (wavenumber versus transmission versus time). In this latter case, netCDF has a facility called *hyperslab*, which would allow a chromatographer to extract a particular two-dimensional array from the three-dimensional data set without having to read in the entire file. This same software could be used to contain an entire data set from a medical imaging experiment, move it from platform to platform, and extract data slices without having to allocate memory for the entire data structure. Sequences of images could be handled as well as a single file with two- or three-dimensional segments removed for examination.

In order to develop a data exchange standard, the AIA Chromatography Standard Subcommittee, working with vendors and users, organized the data into five categories:

- *raw data*—the digitized points of the instrument output including detector sensitivity and scaling information
- *computed results*—peak area, height, retention time, etc.
- *full data processing model*—data processing methods, calibration data, etc.
- *full chemical method*—instrument configuration, instrument methods, conditions
- *GLP information*

As of this writing, only the first two categories have been implemented.

In each implemented segment, the variables and attributes were defined. Those definitions, plus the netCDF storage format constitute the standard. More detailed information, including the details of the definitions and software can be obtained from the Analytical Instrument Association, 225 Reinekers Lane, Suite 625, Alexandria, VA 22314. This information is sufficiently important to the progress in laboratory automation that we've included the specification guide for the AIA Chromatography Standard in the appendix of this book. Perhaps reviewing the specification, and realizing the potential benefits for both vendors and users, will foster the development of data format standards for other techniques and disciplines.

New implementations of standards, such as the upcoming Mass Spectrometry Standard will create a new template (a set of variable and attribute definitions) based on the same underlying software. Beyond that, work is also beginning on Fourier Transform Infrared Spectrometry and the Atomic Spectroscopy Standards. As noted earlier, and repeated here, these efforts are a major turning point in the advancement of laboratory computing. They will have as significant an impact on the way you work as did spreadsheets in business applications.

There is nothing in the structure of the work described that limits its applicability to analytical chemistry. Instrumental techniques for material analysis (stress–strain curves for example) are also likely candidates for data format standards. Clinical data and digitized x-ray can also benefit from similar work.

Within our model, the data format standards address the output of the instrument, and the work shifts from dealing with the physical sample to an electronic representation of it (measurements, properties, etc.). In the knowledge/information/data diagram, we have created the first database—that for *data*. The work now proceeds to using it.

While these standardization efforts are predominately chemistry systems, there is nothing to prevent them from being applied to other scientific disciplines. The same factors and benefits driving analytical chemistry can be found in materials research, clinical work, experimental biology and physiology, physics, and engineering. The same approaches described here can be applied directly to them.

THE DATA LIBRARIAN

The next step in our move through the model is the *Data Librarian*. Most of the introductory material is covered in Chapter 2. Just to recap, the purpose of this element is to manage all the standardized data format packets that come from the various instrument stations. All data is placed in the librarian, and all work on data is based on requests to the librarian. Whether it be a mathematical treatment, interactive graphics, report generation, batch processing of data, or simply a search, it is through that facility. (Note: the librarian doesn't exist yet, but is a needed technology.) The following material provides more detail on the nature of the librarian itself. If we were to create the data librarian as a product or project, these are the basic issues that need to be addressed.

The data librarian will provide a means of storing, retrieving, searching and recording access to laboratory data. In addition, it provides

- long-term archiving of laboratory data—an increasingly significant consideration for regulated industries and product liability cases

- a means of preventing data loss
- the basis for developing a system of third-party analysis and reporting packages
- a required function in the development of integrated laboratory systems
- a significant step in solving the instrument-to-LIMS data connection problem (LIMS refers to Laboratory Information Management Systems)

Within the information flow diagram developed by the Laboratory Automation Standards Foundation (LASF), the librarian package occupies a pivotal position (see Figure 5-8)—the full details of the LASF model are covered in Chapter 2. Laboratory data, collected and formatted by instrument data systems, is exported in files meeting industry standard data structures.

File names are decided by the end-user. Without the librarian, these files would be distributed in directories established by the user. The potential exists for re-use of file names, which would cause the loss of data or

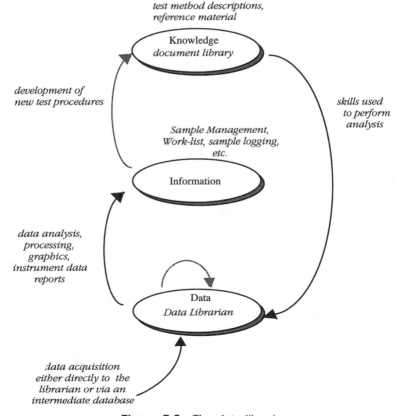

Figure 5-8 The data librarian.

confusion about data content and integrity if files with the same name were found on multiple disks. The contents of files and their names would have to be managed by each user. Unless programming were developed by the users of laboratory data systems, there would not be any mechanism for searching data, aside from a brute-force review of the manually managed file log.

The librarian offers an automatic method of storing and retrieving instrumental data. It would manage the files, provide for searches, distribute the data across media to optimize access and reduce storage cost. Programs, using built-in access controls, can access any data for analysis and reporting. The system would maintain logs of data access, audit trails, and parent-child relationships between files—facilities that are needed to meet regulatory requirements (Good Laboratory Practices [FDA, EPA], current Good Manufacturing Practices [FDA], Good Automated Laboratory Practices [EPA], and ISO 9000). These logs would also provide a chain of custody required to support claims in product liability cases.

Once the librarian is in place four things will happen:

- Laboratories will be in a better position to manage and use valuable data—*solving an increasingly acute need.*
- We can look to the development of libraries of data analysis and reporting packages—*developers and researchers can concentrate on analysis procedures without having to create an entire acquisition-storage-analysis-reporting system.*
- The automatic transfer of data into LIMS systems *without* the need for customized data transfer package will be practical—*reducing the cost of implementing laboratory automation systems.*
- The validation of laboratory systems to meet regulatory requirements will be simplified—*further reducing management and system costs.*

INITIAL CONSIDERATIONS FOR IMPLEMENTING THE LIBRARIAN

Scalability and Growth Paths

In order for the librarian to achieve its full potential, it needs to be scalable—able to manage data collections on PCs (including Macs), workstations, client/server systems, or mainframes.

Database Design

The database design should follow the IEEE Mass Storage System Reference Model.

The intent of the librarian is not to create a system that manages data, but one that manages data files. The librarian does not have to incorporate the laboratory's data into a database, but needs to manage the location of data files, avoiding filename conflicts in the process. Some information about the data files needs to be associated with the master file directory to make searching efficient.

Part of the librarian's function will be the ability to read the data files. Reading administrative information from the files, rather than having the user type it in, will provide for file verification and ease of use. This means that filters will have to be available in a directory that will tell you how to read the data files. The list of filters should be able to be updated without stopping the librarian. The librarian needs to operate in nonstop mode and be able to recover from less-than-graceful shut-downs (power failures, etc.).

Each data file will have an associated history file that travels with it as it migrates from one medium to another. This is a suggestion, not something that is a hard and fast requirement. File usage history (chain of custody, etc.) is important, and it may be easier to do this through a set of individual files than to burden the master directory with linked lists of historical data.

As the library of files grows, media management becomes important, and the system has to manage the migration of files to and from tape, disks, optical media, RAID systems, etc. Unitree (formerly from General Atomics, now part of Open Vision) is a software system that does this. It is based on software produced at Lawrence Livermore National Laboratories.

Backup of the database needs to be considered. It is a significant problem. Small databases can be backed up to off-line media. Storage Technology has a "near-line" system for managing large data files, and IBM has a robotics system. These large systems are capable of managing terabytes of data. Given the amount of data being generated in a laboratory, how do you backup a dynamic system? How do you recover data if backup isn't possible? Some RAID systems have addressed this issue—it needs to be carefully considered in light of regulatory requirements.

The librarian should not be designed with traditional commercial database systems (Sybase, Oracle, etc.). These systems provide a number of tools for programmers to use to access their contents. Those tools can be used to subvert audit trails and history files, and that should not be permitted. Any information within the data system should be accessible to the user in printed or machine-readable form—we are not trying to control the data, just manage it— but steps need to be taken to prevent tampering.

Data entry, search, and requests for data should be possible through an interactive mode or program access. Under program access it is conceivable that a user may point to a directory and say "enter all of the files in that directory into the system".

Security

Assume the worst. The FDA has recently produced a specification for the electronic identification of individuals. The software should incorporate this specification.

INITIAL CONSIDERATIONS FOR IMPLEMENTING THE LIBRARIAN

Librarian Functions

The basic functional model of the data librarian is the circulation desk of a large public library, with the following additions: all transactions must be recorded, people who borrow a file do not get the original data file but a verified copy of it, putting a file back does not replace the original file, but creates a new entry with a pointer to its predecessor. *Note:* files do not have to be returned if no changes were made. Basic functions should include

- *Initial submission of a file*—either manually or through program access to a port. A batch file submission should be possible, and an initial query to the user should provide all necessary, repetitive, or predictable data (counters, prefixes, etc.).
- *Requesting of a copy of a file* or a set of files given a search criteria.
- *Returning a copy of a file*—returning a copy of a file should always be treated as a new submission with the addition of pointers to the original file.
- *Removing files*—this is an interesting problem. In regulated environments files should not be deleted as a normal practice. If they are, a record of who deleted them, when, why, and who authorized the deletion needs to be kept. In some systems—non-regulated research for example—deletion of old or pointless data is normal practice. Regulatory support should not be able to be turned off or on, or overridden. It is either there or it isn't.
- *Catalog searching*—obtaining a list of files with user-supplied search criteria. There may be two modes: searching based on criteria in the master file (should be fast), and, searching based on the contents of data elements within data files or history files (very slow since some data will be off-line).
- *Maintenance functions*—the rest of the stuff I didn't think of. Adding, updating filters, etc.

Other Notes

The basic librarian should be simple. Real user power will come from add-on functions that create and use the data. Those should be separate modules. One example is a "methods" manager used to create and maintain analytical test methods. As far as the librarian is concerned it is just another kind of data.

With this structure, there is no inherent limitation to the scientific market. Data management is a universal problem.

Working with the librarian directly would be a matter of requesting a search based on certain criteria and then getting a list of file names. The application program would have to extract the necessary data from each file to complete its job. That could prove to be a tedious effort. Another approach would be to create an "agent" to do that sorting for you. The search criteria are presented to the agent, and it eventually returns a file to you containing only the data you need to work with. It would be doing exactly the same thing you (or your program) would be doing if it were given the file name list, but that programming doesn't have to be incorporated into every application, just the request to the agents. There are several advantages of this approach over building data extraction features directly into the librarian. First, it separates data management from data processing. Second, it automatically distributes the data processing load. The system containing the agent supplies the processing power: the librarian's function is just moving files. It also means that discipline-specific agents can be created and marketed as add-on modules to the librarian. The agents can be tailored and optimized for particular tasks without touching the code for the librarian, easing the support and validation burden.

Depending on the type of work being done, the data librarian might be the lab management system or it might be a major link in the chain leading to one. In testing laboratories, that lab management system is called LIMS.

LABORATORY INFORMATION MANAGEMENT SYSTEMS

Since the early 1980s this has been one of the more interesting segments of the marketplace (the term was coined by Perkin-Elmer in its LIMS 2000 product, the marketplace would be even more interesting if they had trademarked it). In the knowledge/information/data model it occupies the "I" database. This is not going to be a detailed discussion of the depths of LIMS or how to go about purchasing one. That topic is covered in references such as

> "Laboratory Information Management Systems Workbook", Mike McGinnis and Herman Bailey, Transition Labs Inc., 15020 West 52nd Ave., Golden CO 80401
>
> "Laboratory Information Management Systems", R.D. McDowell, ed.; Sigma Press, Wilmslow, Cheshire, England, 1988
>
> In addition there are numerous papers in *American Laboratory* (International Scientific Communications) and *Chemometrics and Intelligent Laboratory Systems: Laboratory Information Management* (Elsevier).

We are going to look at the strategic issues involved.

The original intent of these database products was to complement instrument manufacturers position on data stations—"data" meant instrument-generated values, everything above that was information. The LIMS was supposed to hold all of the management and descriptive material about a sample: sample identification, source, testing to be done, date required, type of sample, etc. As a management tool it was intended to provide an easy means of keeping track of samples, schedule, workloads, results, and generate reports. And they did that.

The problem with LIMS wasn't the software. The systems available from reputable manufacturers either worked or were made to work. One issue was the name "LIMS". It didn't mean anything in particular, and to some it meant everything. From the vendors' perspective, it was basically a sample tracking, sample management system. Some users saw it that way too. Others saw it as the centerpiece from which everything in the lab was run. That included word processing, graphics, raw data management, anything that had to do with a computer was LIMS. Product managers sweated, developers saw job security (as long as the product wasn't canceled), salespeople smiled, consultants went shopping for a new car. Product managers are responding to the constantly increasing wish-list by expanding product capabilities, creating behemoths in the process, and restricting user options for change in the future. The more desirable systems are built on independent databases (Oracle or other SQL based systems) that provide the possibility for changing vendors in the future and maintaining control over your data.

Another major problem was the lack of consideration given to what a LIMS means in laboratory operations. Managers didn't realize the potential a LIMS offered for restructuring a lab, the misconceptions about capabilities, or how much they had to learn about the way their facilities worked. "I never knew so much about how we worked as I did after we put a LIMS in" is a common comment. The bulk of the issues with LIMS have to do with people, not technology.

From the standpoint of computing strategies and implementation issues, most of what we have been talking about throughout this book brings us to this point. The issue of laboratory information management is a function of data/information usage and flow.

If we ignore the specific products that exist, and use LIMS to mean what people describe when they list their requirements—basically the computer-assisted management of laboratory work—the problem becomes easier to grasp. With respect to Figure 5-9 where do people spend their time and what problems need to be solved?

In a testing laboratory, technicians and entry-level degreed people (those doing routine work) will be mostly concerned with the lower left quadrant of the figure (data acquisition, data storage, and analysis), lightly concerned with the sample management information (primarily sample logging, work-list generation, results entry), and moderately concerned with the reference material for test procedures.

Senior individual contributors (those doing nonroutine work or method development/problem resolution) will spend their time between the top and bottom of the figure. They will be working with the test procedures, instruments, data, protocols, and new or revised test methods. The use of the sample management system will be light.

The laboratory managers will be concerned with the middle, since that is

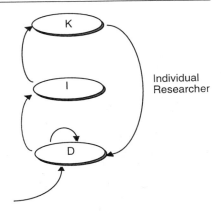

Figure 5-9 K.I.D. diagram.

where the tools that can be used to study performance, work load shifts, lab data quality, costs, etc.

In the course of their work, all three groups will use spreadsheets, word processors, analytical analysis software, and so on—software and hardware that will come from a variety of vendors. Deciding how to implement these modules into LIMS requires careful thought. Do you want the lab to work through a single common user interface or execute software on an as needed basis? This is one of the single most critical determinations you will make. To go back to the woodworking analogy, it's the difference between having separate tools for drilling, cutting, sanding, turning, and planing or one tool that does all of them (there are such things). When you work with several tools, each can be designed to do its job well. The all-in-one products require compromises in design that can be awkward and reduce flexibility.

That decision also has a bearing on implementation design. Some commercial LIMS packages that purport to offer the all-in-one approach require a large central machine that runs everything. Desktop computers can be "used" with them, but their functionality is reduced to a terminal while accessing the central software, and that use will preclude running other time-critical applications at the same time. Truly distributed client-server systems provide more flexibility and better distributions and use of computing power.

Choosing a single product shell for all your work implies a single vendor solution, and less reliance on the labs having to be directly involved with the implementation of an integrated system. In reality it doesn't work. You are going to be involved since no single vendor is likely to provide all your instrument and automation needs. Someone—you—is going to have to provide functional specifications and make implementation decisions. The laboratory manager—and the responsibility can't be delegated—will be responsible for the validation of the system and meeting regulatory requirements (given a global economy, and growing regulations, everyone is going to be required to validate systems—besides systems validation makes good sense even if regulations were not a factor: how else will you know it works?).

LABORATORY INFORMATION MANAGEMENT SYSTEMS

There is considerable advantage to viewing LIMS as a multi-component system rather than a single commercial or in-house developed entity:

- more flexibility in design and implementation
- greater ability to respond to changing requirements and to incorporate new software concepts (data librarian, data processing plug-ins)
- greater control over the data and information produced by your lab
- the ability to phase in systems—once the initial design is in place

Vendors of all-in-one type products will counter that the burden of performance, implementation, upgrading, and so on rests with them and that a single vendor solution offers greater prospects for integration. Vendors cannot relieve you of that responsibility, they can only charge you for it.

From the viewpoint of a LIMS vendor, the laboratory workload looks like that shown in Figure 5-10. The problem is that it ignores sample storage, preparation, data acquisition, and processing, except as they are elements of work lists and method definitions. User requests for automatic data entry from instrument workstations usually lead to consulting projects or quotes for "tailoring", which amount to the same thing. The end result is a unique system, designed to meet your requirements, which now must be maintained

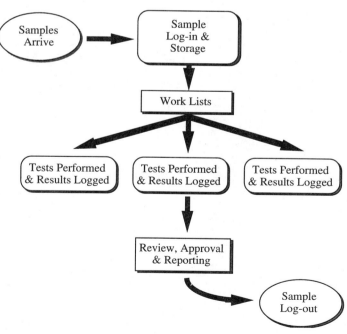

Figure 5-10 Laboratory workload.

as a custom software system. Changes to operating systems, hardware, upgrades to underlying software, etc., are all opportunities to invite the vendor in to resolve incompatibilities and revalidate the system.

This situation exists because, prior to recent work on standards, there was no alternative. Standards-based implementations, which you can specify as part of your strategy and implementation plans, should go far in relieving the current situation and the dependence on custom code.

While these pages, so far, can be viewed as the dark side of LIMS, it's a realistic appraisal of the state-of-the-art to date (ask your vendor for the bright side). LIMS have significantly improved the operations of testing laboratories. They can be a learning experience. At the 1992 LIMS Conference an informal poll was taken (show of hands during a lunch) and fully two-thirds of the people attending were on their second or third LIMS implementation. Reasons for problems in LIMS installations were offered by Robert McDowell in his paper at that conference. First, there needs to be an understanding of what a LIMS will and will not do:

A LIMS will	A LIMS will not
reduce the need to add new people	reduce current headcount
provide a basis for optimizing an organization	fix organizational problems
reduce errors and improve the quality of lab operations	will not exactly fit all your needs out-of-the-box

A LIMS is a change in culture for a laboratory. The way things are done will change, and people may have concerns on how that change will affect them. It is an opportunity to work on organizational issues, but left to themselves, organizational problems will only become magnified by the introduction of one of these systems.

Success depends on

- A commitment to the project by management
- cooperation and support of the people that will use or be affected by the system
- willingness to make decisions when needed

Dr. McDowell's work goes on to look at the causes of failure:

- *Failure to Learn*—from previous implementations, a constructive critical appraisal is needed of the past work of your group as well as that of others

- *Failure to Anticipate*—not taking into account the requirements of other affected organizations
- *Failure to Adapt*—people not taking advantage of the opportunity for change, fear of new methods
- *Catastrophic Failure*—poor technology choices, incompatibilities with other systems, poor project management

From the material covered so far, careful consideration of a computing strategy and implementation plan can improve the likelihood of a successful system design and implementation.

ELECTRONIC LABORATORY NOTEBOOKS

The electronic laboratory notebook (ELNB) is the researchers equivalent to a testing laboratories LIMS. There are some major differences:

- Where LIMS have been in use for over a decade, computer-based notebooks are just beginning to be discussed and have their initial development.
- LIMS systems have a functional model to work against (QC labs have a similar behavior and set of needs). Laboratory notebooks lack this common structure. Real ones, paper, have as much flexibility as paper and pencil permit. There are no limitations on drawing format, table structure, or interchange issues.

There are no electronic notebooks in existence for general use, although some limited application examples exist. Both *Mathematica* (Wolfram Research) and *MathCAD* (Mathsoft, Inc.) have facilities that they identify as notebooks. Test, graphics, and equations can be combined from within the application into a running document.

Among the issues that need to be decided are

- Exactly what is an electronic laboratory notebook? Should it be a machine equivalent to the current paper version or can software and data base technology be used to make significant advances beyond that? Do hyper-information structures have a role?
- How will such a product change the way research is done?
- How will the design reflect regulatory issues (in particular audit trails and validation)? Can a fully electronic system be treated as the equivalent of a paper system (work is being done by the Pharmaceutical Manufacturers Association and the Food & Drug Administration on electronic signatures)?
- And the big issue: How will this kind of system mesh with or conflict

with existing organizational structures? One of the prime lessons to be learned from the simpler problem of LIMS is the need to address organizational and people issues. How organizations really work (versus management theory), what are the short-cuts, and people's willingness to adapt to a machine-based system spending more of their time in front of a tube, all need careful consideration.

Implementing these systems is going to require more attention to backup, off-site storage, and disaster prevention. As more of your company's worth becomes committed to machine-generated and machine-readable media, more effort is going to be needed to protect it. As was noted earlier, people are not as faithful about backup as they should be. Natural disasters (Hurricanes Andrew, Hugo, and Gloria, earthquakes, the 1993 flooding of the mid-West, the 1992 California firestorms) and man-made destruction (1993 World Trade Center bombing) are all reasons to carefully consider the implications of committing your company's wealth to fragile media.

The questions on the technical side are just as significant. Will the ELNB be a working document or a recorder of thoughts and ideas?

In the latter case the recorder becomes a place where information and data are gathered rather than a place where they are used. This greatly simplifies the design of the system; in fact, as will be shown later, a simple prototype already exists.

If it's a real working document, then things become complex quickly. In that case the user will only use one application—the notebook. Everything will be accessed through that medium; it becomes the user interface, database, word processor, etc. In practice the ELNB is an operating system of its own, either layered on top of an existing system or one built for this specific purpose from the ground up. All applications used will have to be integrated into it just as they are with existing systems such as the Macintosh, UNIX, or Windows as its variants; DOS isn't worth considering because of its memory constraints. The reason that it needs to be considered from this standpoint is the regulatory requirement for an audit trail, which in turn reflects reasonable laboratory management practices. If left to each product, the risk of incompatibilities is high and the behaviors will vary. You may wind up with multiple audit trails, one for each product and one for the notebook. The best way to enforce an audit trail is from within the operating systems itself; that would provide for standardized routines, information content, and file structures. Doing it at the O/S level also makes it easier to implement newer technologies. Hyper-information structures (which will be covered later) could be included more effectively than if left to individual applications. In addition, there are these points:

- Will we be able to integrate existing products as is or will they have to be redesigned to fit?

ELECTRONIC LABORATORY NOTEBOOKS

- How will we accommodate multi-platform requirements for file/document sharing and distribution (proprietary implementations on single platforms aren't worth considering; you yield control of your data to the vendor)?
- Will the ELNB be scalable? Can we have a small version for a single desktop system or will a network with storage servers be mandatory?
- Finally, will an electronic lab notebook be a product or project, requiring extensive consulting to get it running and keeping it functional?

The storage requirements for this type of application are gong to be huge. Client-server approaches, coupled with intelligent file management systems that automatically move files between media types depending on use may be mandatory. As a result, you may not be able to take your notebook home with you, at least with today's technology (the future has a long way to go, and there are some very bright people out there). One of the things computer applications have in common is that the files they need to work with must be immediately accessible to them. Suppose we have a system with all the bells and whistles, fully integrated applications, audit trails, hyper-information links, and a built-in coffee maker. Maybe you are using disk structures that allow multiple disks to be used as a single volume. Eventually they will get filled up. If information on page 47 is linked to data on page 2, which might be linked to someone else's notebook, how do you off-load files? You can't without breaking links and losing the benefits that the system was designed to produce. With linkages possible, you may not be aware of the extent of the linking network. Will you in fact ever be able to off-load anything? Intelligent file management systems can handle the issue, but that limits portability and effective use.

The ELNB as a recorder is a much simpler problem. The notebook becomes a place to accumulate material rather than a place to generate and manage it. The development of data and information would be done through existing applications. With the exception of the ability to provide automatic (enforced) audit trails, the tools to begin experimenting with such a system exist today. The criteria we will be considering are these: the ability to place graphics and text, generate a table of contents and index, and produce a printed version with page numbers, and a template for signatures. Literally an electronic version of the paper system, but with higher quality output, more flexibility and easier management. The description below is not of a fully functional production ELNB but a means of experimenting with the ideas to get a better appreciation of what we really want.

Those requirements are met in page layout software used by graphics designers today, the best are Quark Express (Quark Inc.) and Pagemaker (Aldus). The process described below could be done with a high-end word processor; the major limitations are the reduced flexibility in placing graphics and text and the ability to product text that is not linked (and, as a result, will not flow and alter text and graphics placement).

Begin by creating a template for the working pages of your document. This would contain any headers (with optional company logo) and footers including place holders for automatic page number placement and a signature block. Hold a few pages in the beginning of each document for a table of contents and at the end for an index. Both systems support adding pages as needed so the initial space allocation can be small.

From here on, it's a matter of using it. Text can be entered from a word processor or created using the software's internal editor. Graphics and tables can be imported from a variety of sources (spreadsheets, drawing packages, slide presentations, etc.) and placed. Captions and annotations can be added. Audit trails become a matter of manual annotation. Both Quark Express and Pagemaker support generating a table of contents and index so material can be easier to locate. Pages can be printed on a regular basis, signed, and kept in hardcopy form for regulatory purposes. Versions can be distributed in essentially read-only form through third-party page capture software. Output from most software packages can be imported, so that is a plus.

While simple-minded, this approach can work and does begin to shed light on potential problems that will occur in a real ELNB; for example, file management. Both page layout packages will permit you to either make a graphic element part of the document or have it remain as a separate file with a one-directional pointer to the graphic file. For small objects it's easier to include it; for large one you may want to have it reside separately to reduce disk storage.

If you delete the graphics file, the page layout software will only have a low-quality copy within it and you won't be able to print it. You can change the name and lose it as far as the document is concerned, or substitute another for the original with the same name and get a mismatch between what you thought was there and reality. Always importing the graphics into the document solves that problem, but the document becomes large quickly and performance may suffer. If you want to use disk space quickly, start importing images, animation, and QuickTime (Apple's video software standard) movie clips, interesting ideas for a notebook.

Both the Windows and Macintosh versions of these products offer dynamic updating of data, so changes can be incorporated easily. *But* since these automatic updates do not provide an automatic audit trail, that will have to be provided manually. There is also the concern that changes will be made through these links inadvertently and not be properly audited, so use Publish and Subscribe (Mac) or Dynamic Data Exchange (Windows) with extreme care, or not at all. This is only recommended as a learning exercise.

So far the material on lab notebooks has been viewed as a single-user environment. A lot of work is being done in groupware, multiple interactive users on different computers. There are a growing number of applications offering the ability to have several people working on the same document at the same time. This can be an interesting exercise in group dynamics.

Although groupware has been under investigation for several years (interactive systems were demonstrated at Xerox PARC several years ago), they haven't been operating as commercial systems long enough to get a good understanding of their strengths and weaknesses. Product development is progressing on groupware. An article by Paula Rooney (*PC Week*, August 23, 1993, p. 8) details the following points:

- Microsoft is rumored to be in discussions with Future Labs, to produce a product called "Whiteboard", which will let multiple users modify the same document at the same time.
- Electronic Meeting Systems, Inc. (Cambridge, MA) is getting ready to release a product that will let users integrate word-processing, spreadsheets etc., on screen. Future versions will include time-stamping.
- Conferencing technology is being incorporated in products by Future Lab, Xerox, and Fujitsu Networks Industry, Inc.

Technical issues (performance, network traffic, etc.) aside, the biggest issues are people. Think back to the last meeting when a group (at least 3 people) had to work on a document under a deadline pressure. How did it go? How important was face-to-face contact? Would having a computer between you have helped or hindered the process? Suppose it was the only vehicle of communication? Remember, the point of a computer is to help people work and to change the work where that change is beneficial—it's a tool.

Seriously considering an electronic laboratory notebook that includes dynamic access will be a challenge to both software engineers (remember the need for audit trails) and those in group dynamics. People's attitudes about ownership of data, security, and egos will come into play. Large scale file storage, backup, disaster recovery, proper management of corporate assets, the real value of data and information all have to be considered with care: this is a fundamental strategic issue.

The next Chapter will review computing technologies.

CHAPTER 6

Computing and Information Technologies for the Laboratory

In Chapter 5 we moved through the laboratory model and looked at what had to be done in each area from a strategic viewpoint. This chapter focuses on underlying technologies.

Figure 6-1 was discussed in the section on implementation. It is repeated here to relate the material covered earlier to what is going to follow. Each of the applications technologies (LIMS, instrument data systems, etc.) is built upon computing technologies—the focus of the material below. The intent is not to provide a thorough analysis of each technology, but rather to acquaint you with their possibilities, capabilities and issues, so that your computing strategy can take advantage of them.

There is no doubt that computing technology has transformed laboratory work. Much of what we take as routine today would not be possible without the equipment provided by it. The portion that would be possible using

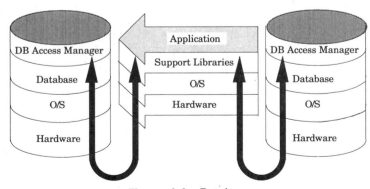

Figure 6-1 Databases.

manual procedures would be prohibitively expensive, assuming we could find enough trained people to do the work. The changes we have seen are the beginning of an accelerating process of change.

While much is possible, and vendors will stress lots of possibilities, there are questions of what should be done. What you *can* do with technology and what you *should* do in your laboratories are matters for careful consideration. Again it's a matter of developing an informed information management and computing strategy.

Note: Throughout the following material the word "system" will be used frequently; you can't avoid it—the fingers start to shake. That word can be interpreted as "environment", or, as a collection of "stuff". There is nothing magical or mystical about that word.

OPERATING SYSTEMS

If the applications you are working with haven't forced a choice, then the operating system you choose to work with is one of the most critical decisions in implementing a laboratory computing strategy. The operating system (OS) will shape the computing hardware used and place limits on what you can do and at what expense. The cost of services (end-user support, maintenance, installation, training) will all be a function of that initial decision. This is not going to be a tutorial on operating system design, structure, or how to use DOS. Book stores are full of them (the local bookshop has just expanded its computing section, sadly at the expense of the general science section, to five racks of material, each about 5 by 15 feet). This is going to be a review of the strategic issues.

Operating systems can be grouped by overall capability (ignoring network access):

- **Single Task:** Only one program [task] can operate at a time. Only one person has complete access to the system, and only one program is running at a time (examples: MS-DOS, early Macintosh systems [Finder, pre-System 7], CP/M). *Note:* In current versions of operating systems, the distinction between single-task and multi-tasking systems is blurred. Early single-task systems were purely that, one thing at a time; today's systems may have only one user program running (such as a word processor), but other "tasks" (such as print spoolers) can run at the same time. Those tasks are usually of very limited functionality. Best use: time-critical applications where guaranteed system response and freedom from interferences are mandatory.

- **Multi-Task:** Several programs are executing at the same time, sharing access to the CPU on a priority basis. Programs may communicate with each other, share files, or operate without regard to other programs (examples: VMS, UNIX, RSX-11M, Windows 3.1, Macintosh [MultiFinder, System 7]) OS/2. Best use: time-critical applications where predictable system response is important, and several independent or closely cooperating tasks are running at the same time. The system response will be more predictable than having several computers cooperating due to varying network loads.

- **Time-Sharing:** Several users have access to the computer at the same time, with file-sharing, messaging, etc. Each operates at approximately the same priority, and the CPU is shared by giving each user a portion of its time (examples: VMS, UNIX) Best use: large database applications where system security and timely access to updated information is critical.

The OS is a software package, it can be viewed either as a single large program with a large number of supporting modules or as a collection of programs. Different OS's from different vendors will fit one description more than an another. The tendency over the past few years has been toward the former (Windows and the Macintosh for example) and away from the latter (CP/M, DOS). We are going to begin by taking a look at the development of operating systems as they emerge from the primeval substrate of hardware. How we got were we are today is important, since that history has shaped the systems we have on our desk today and will explain why as a community we have some serious choices ahead of us.

Historical Digression

This is not going to be THE history of computer operating systems, just a review of what has transpired in the recent past.

When computer vendors first began selling machines, their object was to sell hardware: CPU boxes, printers, terminals, etc. The concept of software as an independently marketable product was not taken seriously until the late 1970s and not put into practice until the early 1980s. Computers were sold to people who programmed them to do something useful. Minimal operating systems were provided (almost free of charge in some cases) as a means of making the hardware easier to use and program.

The earliest "operating systems" were people. The computer system consisted of a CPU box, printer, and keyboard, along with a papertape reader, card reader, or magnetic tape drive. Files consisted of tapes (paper or magnetic) or decks of cards; you were the file manager. File storage consisted of boxes, drawers, or whatever was empty. On early PDP-8s (12-bit word length, usually 4 to 8 kilobytes of memory, Intel 80386's are 32-bits and usually start at 2 to 4 megabytes of memory) the programmer would load a short program into memory—the RIM loader, about 14 instructions—which

OPERATING SYSTEMS

would read a more sophisticated program into memory. That program, the binary loader, had checksum error checking. The binary loader was used to load your program and off you went. So the world ran until the late 1960s (some systems like this were in productive use in the mid-1970s).

Disks having capacities of anywhere from 32 kilobytes to a mind-boggling 2.4 megabyte came on the scene, and people wanted to use them. So was born the first Disk Operating System (DOS, and there were a lot of them, every vendor had a least one). The primary purpose of a DOS from the vendors view was to make the hardware easier to use and hence to sell. Those DOS packages had a few simple characteristics including

FEATURE	EARLY DOS	MS-DOS
File storage	Files could stored and referenced by (on PDP-11s, 16-bit machine) a 6-character name with a 3-character extension no subdirectories	Subdirectories allowed, file names can have 8 characters and a 3-character extension
File maintenance	Create, delete, copy	File and subdirectory create, delete, copy
Program loading	Yes, via load or run command	Yes, type name of program
User written device drives	Yes	Yes
Type of OS	Single task	Single task with small terminate and stay ready routines

Functionally the DOS systems of today and 20 years ago are very similar. The costs of the hardware are two orders of magnitude apart, and memory and disk capacities are separated by three orders of magnitude.

Since the price of computer systems with disks, CPU, memory, terminals, printers, etc., easily ran $50,000 to $100,000, ways had to be found to make their use more attractive. Memory capacities could be increased to 128 kilobytes, and then to 1 MB. Disk capacities increased to perhaps 5–20 MB. Operating systems changed to meet user requirements for doing more work on a single machine and to meet competitive pressures. Price and how easy it was to program were the key factors. Time-sharing operating systems were developed to allow more people access to the same computer and thus reduce the cost of hardware per user. Time-sharing was possible since the user interface was minimal (similar to a typical DOS screen, ignoring DOSSHELL) and people were much slower in reaction time than computers. Many users could peck away at keyboards while the CPU serviced them and its peripheral—there was only 1 CPU, no smart controllers.

Another factor that made time-sharing acceptable was that people were willing to wait when the CPU got very busy. There were no real-time or time-critical events, so if placing a character on the screen took a second or so longer, no value was lost. An "A" is an "A" now or 10 seconds from now. As time-sharing became an application of computers, vendors responded with faster CPUs, more memory, faster disks, higher capacity disks, and so on.

Time-critical computing was becoming a factor in laboratory and process control work. Time-sharing operating systems could not do the job. Multi-tasking operating systems with fast response to interrupts (signals that indicated that something needed immediate attention) were developed. These would allow several tasks (with one or more users) to operate at different levels of priority. Time-critical events had high priorities, getting computing resources when they needed them, and those that were not got to share whatever CPU time was left over. Since the user interface was minimal, the overhead per user was nil (today, graphical interfaces account for a significant fraction of the CPU's work load). This development led to process-control computers and laboratory systems capable of servicing 16 chromatographs and 4–8 users on the same machine all in 32KB of memory.

This multi-tasking capability exists today in Windows and the Macintosh System 7, but in a rudimentary fashion. We have, today, the ability of having several tasks in various stages of execution at the same time and doing network support, printing, and some copying at the same time. What we don't have is the ability to handle time-critical interrupts on a priority basis; it is possible for a word processor or screen-saver to inhibit data acquisition. Since the bulk of computer systems are sold to people using "business" applications, the trade-off between the complexity of priority arbitration and real-time response against ease of use was a simple choice.

The need for sharing systems due to economic issues is past. It is easier and cheaper to manage several desktop PCs than it is to install a time-sharing system. While the cycle of computing has gone full circle and in a sense got us back to where we started on DOS systems (although at a substantially lower cost), that cycle isn't going to continue with the development of new multi-tasking and time-sharing operating systems. The economics of implementation and flexibility are going to favor large numbers of cooperating computers with very few (one time-critical task, or 3–4 non–time-critical cooperating applications) tasks per processor. As we learn to implement large databases on client/server systems, and commercial software migrates in that direction, time-sharing systems will yield to shared access databases and messaging between applications.

There are three choices for operating systems in the laboratory today: single-user (DOS, MAC, UNIX, VMS, etc., depending on the applications need for resources—I'm avoiding Windows NT since not enough is known about its use in laboratory applications), time-sharing, and client/server systems.

Data acquisition, analysis, word processing, etc., are usually left to desktop

OPERATING SYSTEMS

applications. Large database problems, such as LIMS, reside on time-sharing systems though over time they will be shifting to client/server systems. Client/server applications in the laboratory, true distributed applications (vs. multiple people having access to a larger file storage device) have yet to appear although they are under development. There are LIMS applications being worked on, and, if electronic laboratory notebooks become commercial, they will be on these systems.

The critical issues revolve around how we respond to the developments taking place in operating systems. The problems come in two areas: the lack of application of commercial OS's to time-critical requirements and the lack of stability of the software.

A stripped-down DOS system or Windows systems can be used in real-time work today; it is being done on a regular basis. That capability is diminishing over time as more features are loaded into these systems, as their stability becomes troublesome due to poor design or incompatibilities with existing software, and as support for old versions is discontinued.

The operating systems wars between various vendors vying for market dominance is going to favor the business user; there are a lot more of them, and their needs are simpler to satisfy with flashier products. Designing efficient time-critical systems is a lot of work, and laboratory systems don't produce eye-catching displays comparable to a presentation package with QuickTime movies. We have several choices open to us. We can exchange information on how to work around new features that get in the way of lab work and how to dedicate systems to real-time work with non–time-critical tasks being placed on other machines. We can look to vendors to implement data acquisition systems that contain their own OS's, obviating the need for MS-DOS. Or we can remove the data acquisition function from the PC host and put in a separate box. That box would get its programming from the host and report data back in a standardized format.

I favor the latter development. It is easier to control and provides predictable behavior. Current implementations of data acquisition, analysis, storage, and reporting on a single machine are a carryover from earlier generations of systems when communications were slow and cooperating systems expensive. Systems design and validation would be easier with modularized systems. This approach also makes it easier to adapt to changing OS conditions and take advantage of new developments. It avoids having applications depend on out-of-date versions of operating environments.

This method has been used in the past. Two Digital Equipment Corp. product of the 1970s—RSX-11M and RT-11—had the ability to create and download independent time-critical systems to dedicated computers. RSX-11S systems could be loaded from an RSX-11M host, or from media, and run without a hard disk to carry out work and report data over a network. The RT-11/Remote product provided the same capability for a smaller, less capable system. The reason for creating these products was to reduce the cost of hardware configurations (mass storage was expensive); today disks are low-cost, but the same type of solution provides isolation from the peculiarities of commercial operating systems.

Requirements for high speed/high data volumes could be met with the same type of structure. The "box" would contain an A/D control lines, a CPU memory, and a hard disk. Parameter files could be sent over the network, and files could be returned in a standardized data format. As an alternative, the "box" could be a board on standard but configured to look like a peripheral disk. It would operate independently of the CPU taking its direction from a configuration file and store data on its own on-board disk until it is copied to the host system. Zymark uses this pseudo-disk approach to transfer data between the Benchmate and a PC. Boards of similar structure are used as smart SCSI controllers, as signal processors, or as storage and memory upgrades (Kingston Technology Corporations DataCard Micro Channel board for example).

The most troublesome aspect of the current OS competition (which should bring out the best in products) is the lack of stability in the choices available. Apple had its share of problems with the initial release of System 6. Microsoft has had its problems in the past with earlier Windows offerings and is now in the midst of a new round of MS-DOS problems with version 6. The major issue there is the disk compression utility. As reported in *InfoWorld* and *PCWeek* during the spring and summer of 1993, this could cause the loss of all data on a hard disk as a result of installation problems or loss of synchronization between input and output buffers during use. Windows NT has yet to be tested in laboratory work (an article in the August 30th, 1993, issue of *PCWeek* by Aaron Goldberg, p. 118, suggests a problem-free test period of 180 days or more before accepting NT as a usable production system).

If the current crop of OS vendors want to do their customers a real service, they should spend more time on building in product quality and stability. Loading in new features to provide a longer list of items in a brochure isn't the goal. Providing features that have real value, that work, and that are well tested before they hit the marketplace is more important. OS's are purchased as a basis on which to do work; that basis needs to be solid and reliable or the entire structure will collapse. Longer times between OS releases are desirable to allow for stability to develop in layered applications. The installed base of users should not be used as an extension of beta testing. In other markets, dominant vendors have lost market share because of poor quality, leaving that as a wedge for new competitors to gain an opening.

While long test-and-try-out periods are desirable, individual companies rarely have that luxury in resources. Users want the latest and greatest, and vendors terminate support on older versions soon after new ones are released. Incompatibilities frequently exist between the applications you want to use and new versions of OS's. The layering of enhancement on operating systems only contributes to the issue. Someone using Windows has the option of several desktop enhancements riding on Microsoft's product, which in turn could be riding on top any of three vendor versions of DOS! And if something

OPERATING SYSTEMS

doesn't work, you have to figure out what's broken before technical support can be of any value.

While you can't do much about the thrashing in the marketplace, aside from writing letters to corporate presidents, you can make strategic decisions about what products are allowed into your laboratory and under what circumstances. Minimize the number of layered products used. Delay purchases of upgrades until the product has had some real-world use and bugs have been detected and fixed. Check trade publications for reports of problems. Avoid unnecessary enhancements or modifications including the tips found in trade publications. They may be neat tricks, but, given the potential for side-effects, they should be treated carefully.

The vendors' position is easy to understand. The marketplace is dynamic, and a delay of a few months could eliminate any competitive advantage. Given the consequences of bugs and flaws in the software, vendors should spend more time being concerned with product quality and stability. The marketplace should not be considered an extension of beta testing. Flashy new features may impress managers and fellow marketers, but corrupted systems may cost you your research data. Microsoft's DiskDouble problems are a good example. The feature sounds interesting and it may be a "hot" subject, but the trade-off for you is this: do you use a software algorithm to increase disk storage, with the loss of some performance and the potential risk of your data, or do you just buy a bigger disk?

Make sure your MIS department understands your computing strategy and, preferably, was a partner in its development. Since change is constant, you're going to have to continue to adapt, and part of that adaption is protecting your investment in data. Talk to your critical vendors, either directly or through users groups. Make sure they understand your issues. The future of computing is not a matter of evolution—it is a matter of design.

There won't be any clear winner in the OS contest, nor are we likely to see an end to it. The push on OS's is for the general-purpose marketplace. Even there, through primary vendor and third-party add-ons, OS's are customized to solve particular problems—the decision to include QuickTime, for example, is a tailoring of the general-purpose systems. The addition of networking is a modification of the OS. As application needs develop—the development of an electronic laboratory notebook for example—market specific needs will develop that will lead to the tailoring of OS's (either through ground-up design or add-ons) for particular applications.

The underlying hardware is changing, and the capabilities of the operating systems are going to change, slowly, to keep up with them. 32-bit computers have been with us for several years, and we are working with operating systems designed for 16-bit machines (the Macintosh and VMS are exceptions). There is also an awful lot of stuff being packed into today's operating environments: graphics, animation, video, audio, broadcast television. Science-fiction writers may have a point in a direct human–computer interface instead of moving data representations through the senses. The machine on your desk is your gateway into computing and information technology, an entry point but not the entire structure.

A friend of mine once said that there are three things you can't have too much of when flying a plane: fuel in your tanks, air under your wings, and runway in front of you. In computing technology there are a few things you can't have too much of as well: computing power, data storage, memory, and network speed. How do we get them?

DATA STORAGE

Even though disk capacities are growing, there is a limit to how much stuff you want to put on disks within your computer—for no other reason than the responsibility of backing it up, and the risk of losing it through malfunctions, mistakes, or security breaches. That means putting it somewhere else where you and anyone else who needs it can get to it. The answer in today's technology is a file server, a computer with large disk storage that is accessible to anyone with the authorization to get at it—a network is assumed.

File Servers

In its simplest implementation a file server is a relatively small computer with large disks. They may be a series of individual drives, each of which is independently accessible, or an array of disks that operate as though they were one large disk. This is where data files are stored when they are no longer active. In a large lab this is where the Data Librarian (when built) would actually keep its data files and manage them. VMS users with Pathworks can use a VAX disks for storing PC files in their native format and retrieving them through what appears to be a local hard drive. Other vendors, including Apple and Microsoft, provide server support. There are a lot of options, but with that range of mix-and-match products comes the responsibility for integration and support—keep it as simple as possible. Apple's System 7 has the facilities for file sharing and file server support built into the OS, and it works rather well. If the file server is managed by the IS group, be sure to confirm backup procedures with them.

One useful technology for file servers is the Redundant Array of Inexpensive Disks or RAID. You can buy a single large-capacity drive, or, using RAID technology, purchase and equivalent capacity through a large number of smaller capacity drives. Having multiple drives operating as a single logical disk offers an increase in speed and dependability. If one drive fails, its data can be recovered and placed on a healthy drive. The failed unit can be replaced at lower cost. Another advantage is the ability to increase the amount of storage over time in increments instead of making a large financial

DATA STORAGE

commitment at the outset. The entire set of disks is managed through software designed for the task rather than just using the standard disk driver software and a hardware controller.

The next step in complexity is to move all applications to the file server in addition to the data, leaving the local PC's hard disk for the operating system, active files, and temporary storage space. This approach minimizes the need for large local disks, forces everyone to use the same version of a software applications—they are all copying from the same image—and reduces system maintenance cost (multi-user licenses are required from commercial vendors to implement this). The cost reduction comes from only having to update a single copy of the software instead of each user's system. The disadvantage is the heavier network load, delays in applications startup, and the risk (though small) of dependence on a single machine's operations for the successful operation of your facility. If all of this is done in a local environment—same immediate geography—and the network capacity is well planned, this can work well. Problems occur when getting overwhelmed by what is "possible" get in the way of what makes sense.

The combination of fast network capability, file servers, and network access to data can lead to technically feasible (able to be done based on product specifications and capabilities), but functionally unworkable situations. A laboratory management system (LIMS or laboratory notebook) can work well in a distributed environment. How extensively distributed that environment is can make the system fail. A LIMS, for example, installed in one lab may be able to support additional laboratories as well, making the cost more attractive. If those labs are close enough to support high-speed links, the system may work well. As you start to add facilities (direct coupling of instruments and users), network traffic will go up rapidly and may swamp the systems capacity. Poor response time is quick way to generate user dissatisfaction and cause a system to fail. Should the labs be spread over a large area, separated by miles, the system will become unworkable in practice. Once facilities get separated by large distances, the network linkages carry not only laboratory traffic but other corporate traffic as well. In addition to network load considerations, there are practical issues of reliability due to power outages and disruption due to damage to lines, weather, and other issues. Reliability should have a higher priority than capital expenses; the costs due to poor response will quickly overcome initial savings.

Other issues in data storage technology are managing the growth in the number and size of the files and managing the space required to maintain them. Since, at least in quality control labs, you can't discard data (due to regulatory or product liability issues), storage space has to be allocated for it. Smaller laboratories may be able to work using optical disks or magnetic tape as off-line storage—someone will have to act as an archivist to keep track of what is where. (*Note*: The shelf life of magnetic tape is uncertain; tapes should be copied at least once a year to maintain data integrity and prevent blocking of the tape.)

Hierarchical Storage Systems

Larger facilities may need to rely on hierarchical storage systems to maintain data. There are different implementations of these systems, but, through a combination of hardware and software, they do the same thing. They migrate files among different media (disks, tape, etc.) depending on usage so that frequently used files are on the most accessible media and those of occasional use are on lower speed, lower cost, higher capacity devices. Although the media may be inexpensive, the mechanisms that support them may run to ten or hundreds of thousands of dollars—unless you're considering terabytes of storage, that shouldn't be a factor. The benefits of these systems are that they relieve you of having to consider what files to put where on regular basis and many of then provide automatic backup. In addition, the entire collection of disks, tape drives, and whatever is allocated to the systems domain (i.e., you've told it what it can use and what it can't—domain is more impressive) can be treated as one large storage device, so accessing information is a lot easier.

As with anything big, interesting, and expensive, models are useful to understand the underlying structure, and this is no exception. The IEEE-Mass Storage System Reference Model (IEEE-MSSRM—see "A Reference Model for Mass Storage Systems", by Stephen, W. Miller, *Advances in Computers*, Vol. 27, 1988) works like this: let's suppose you have a file that you want to submit to a product that supports the storage model. You send the file, and the following things happen (see Figure 6-2):

- The Filename you use is mapped to a unique identifier within the system.

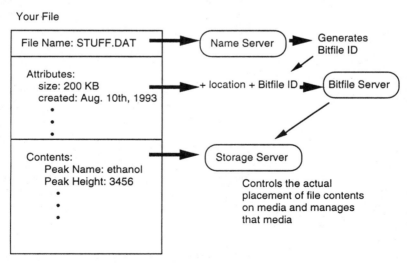

Figure 6-2 Storage model.

DATA STORAGE

Internally, that identifier is used to refer to the file contents. This mechanism is needed to accommodate different operating systems with different naming conventions. Names are managed by a *Name Server*. Your data file is referred to as a *Bitfile* since the contents are treated as a collection of bits—the storage system isn't interested in the contents (just the file's attributes such as name, file ID, size, access mode, security level, where the file is at a particular time, etc.) but only anything that affects the management of the file as an object without regard to what's in it. The elements are referred to as "descriptors" and managed by a *Bitfile Server*.

- When a file is moved from one location to another, that movement, including reading and writing, and network access are controlled by a *Bitfile Mover*.

- Once within the storage system, accessing a file is done as you would any other file. When the system detects that the data is on a storage server, those operating system functions that normally open, read, write, and close a file are redirected to the server software that accomplishes the tasks, but they are directed through operating system's own mechanisms.

There are products on the market that address these requirements. Some are hardware storage systems that require supervisory software to complete the picture, and some are software packages. Conner Peripherals (San Jose, CA) has recently announced a hierarchical storage system. Open Vision (San Diego, CA) has purchased Disco's Unitree and extended it beyond the Unix world to include PCs and other platforms (see "Implementing A High-Performance Hierarchical File System Using Networked Disks", Michael Hardy, *SuperComputing Review*, April 1990, pp. 49–53; the article's references include additional articles on subject). Before we get to the software, there are two hardware issues to be looked at: communications and specialized storage systems.

If your company is managing or considering managing enough data to justify systems like this, then spend some time looking at the network requirements. Plan on the projected needs several years out, then double them. Short-changing yourself on network capacity is going to lead to user frustration, lost time, lower productivity, and poorer morale. Think about the prospect of demonstrating the system to the board of directors and having them stand there while you wait for the system to respond (do you know the feeling you get when you can't tell if the system crashed or is just taking a while to get back to you?). The cost may look prohibitive, but you pay that initial outlay once. Your pay for poor performance every day.

Most of us are familiar with data storage in two forms: it's either on-line and available or off-line and inaccessible until it is mounted (a tape for example). There is also a realm between these two. It is offered by Storage Technology, which they refer to as "nearline". The concept is important because it fits well with the storage model described above and provides a price/performance compromise that is worth considering (*Note*: Similar technologies are available from other vendors). The system provides a mean of automated tape mounting and storage. Other non-commercial examples of this exist in laboratories. CERN (the Center for European Nuclear Research, Geneva, Switzerland) has robots that travel on rails between racks of magnetic tapes which can be automatically mounted and made available for use.

Each StorageTek module consists of a 12-sided cylinder containing 6000 tape cartridges (IBM 3480 compatible) with 200 megabytes of storage per cartridge or 1200 gigabytes total per unit. Several units can be put together to form a cooperative working system—tapes can be passed from one system another. Tapes line the inner walls of the device and a pair of robot arms identify, retrieve, and replace cartridges (each has a barcode to confirm identification)—see Figure 6-3. Each unit has several mounting ports, and if the ports on one unit are in use, the robots pass the cartridges from one to another.

There are other vendors of robotic tape–library management systems. International Business Machines (IBM) has released the Tape Librarian Dataserver to meet this need. The machine is capable of storing 15.4 trillion bytes of storage in 18,900 tape cartridges. The largest configuration is 8-feet wide, 8-feet tall, and up to 92-feet long. The tapes are accessed by a single robot running on a track. A camera-mounted hand fetches the tape and loads it into a reader.

Hardware like this, coupled with mass storage management software that fits the IEEE model, creates a powerful package for solving mass storage problems. The systems can be scalable (starting small and growing as needed) provided the storage management software is introduced early. Doing it late

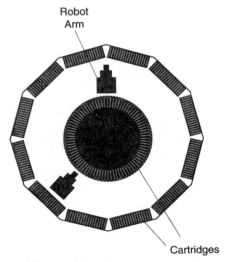

Figure 6-3 StorageTek module.

DATA STORAGE

in the game is feasible, but you spend a lot of time bringing the systems up, tuning them, and integrating them into the system, with a backlog of work waiting. The next step is to look at a software system for mass storage management.

Unitree was initially a UNIX-based product, but it has expanded to support a variety of operating systems, including those provided by IBM, Amdahl, Sun, Convex, Digital, Fujitsu, NEC, Silicon Graphics, and Hewlett-Packard (contact the vendor for details)—things change. Combined with support for disk drives, tapes, and the StorageTek equipment it yields the scalable support need for a large laboratory or corporate mass storage management system that can be phased in as a function of need (see Figure 6-4).

Unitree implements the IEEE-MSSRM and provides for complete data management. Files automatically migrate from one medium to another depending on demand. The system provides for data security, recovery, and backup. Media usage is monitored and consolidated.

The matter of large-scale data storage is an active area of research. A paper ("Storage Systems for National Information Assets, A collaborative Research Project" by Robert A. Coyne, Harry Hulen, and Richard Watson—a copy of the

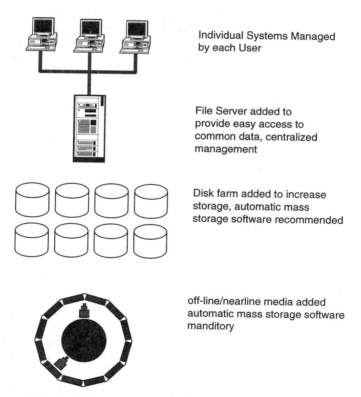

Figure 6-4 Scalable support for mass storage management system.

paper was provided by Open Vision, San Diego, CA) details the current state of a collaborative effort on the part of IBM Federal Systems Company, Ampex Recording Systems Corporation, General Atomics DISCOS Division [*Note*: Unitree is a commercial version of a project at Lawrence Livermore National Laboratories that originated at Discos and was taken over by Open Vision], IBM Storage Systems Product Division, Maximum Strategy Corporation, Network Systems Corporation, and Zitel Corporation. The purpose of this project is to investigate the technology required to meet the national public and private data storage requirements. Among the subjects under study are media types, automatic management of distributed systems, types of access needed to storage to improve performance, and network system requirements to support the access to and movement of large quantities of information. The IEEE holds an annual symposium—the proceedings of the 12th was issued in 1993 by the IEEE Computer Society Press, Los Alamitos, California—on Mass Storage Systems. The contents of that volume contain papers on requirements, storage system standards working groups (*Note*: The IEEE-MSSRM is at revision 5), system architectures, data access, and emerging technologies. A detailed discussion is beyond the scope of this book. If you are interested in the subject, the IEEE series is a good place to begin.

While it may be some time before a lab system alone can justify the more expensive items, being aware of the options open and how they can be applied is the key to planning. If your company is planning on a large-scale storage system it might be worthwhile to consider its capabilities and how the lab can make use of them. *Note*: If your laboratory is considering maintaining digitized images of photographs in machine-readable form, storage may become an issue rather quickly. Even gray-scale images of typical photos can exceed one megabyte per image. While compression schemes are available for images, lossless techniques do not save more than a few percent. Image compression techniques used by graphics artists (JPEG), while very effective at reducing storage space, do at the cost of image fidelity and do not meet the requirement of regulatory agencies.

CLIENT/SERVER SYSTEMS

Client/server "technology" is an encapsulation of what we have just covered. It's a means of achieving distributed access to stored material (data, information, etc.) and services that are needed on a broader basis than a single user. Time-sharing was a means of providing distributed access to a computer system. Client/server systems are a means of providing distributed access to services. The *client* is whoever (or whatever—it could be a program) has access; the *server* is whatever provides the needed service: access to a data file, specialized printing, high-performance computing (a supercomputer), administrative functions, etc. When all the jargon is tripped away, it's that simple. The keys to the successful implementation of these systems are

CLIENT/SERVER SYSTEMS

twofold: a networking system that provides reliable, high speed (qualitative, since "high speed" depends on the application and size of the network) communications and a well-engineered application that balances the processing and user interface load between your desktop system (client) and the system providing whatever it is you need. *Note*: Some LIMS may try to pass themselves off as client/server systems by saying you can use a PC to gain access to the LIMS menu. If they do that using a PC-driven user-interface and query structure in which the PC actually off-loads some of the processing from the central LIMS system, they are correct. If all you are doing is running a terminal emulator, your PC has been reduced to a terminal and all the load is on the host system (the PC is providing the user interface, but so can a dumb CRT in that case).

Client/server systems (see Figure 6-5) become valuable when (1) a collection of information or some service (fast number-crunching) is needed by more than one individual at a time, and (2) when the local computer (PC—the client) can provide a useful function in the process. Consider the following situation: the contents of a database are needed by several people in the laboratory. The database is updated periodically, and the contents can

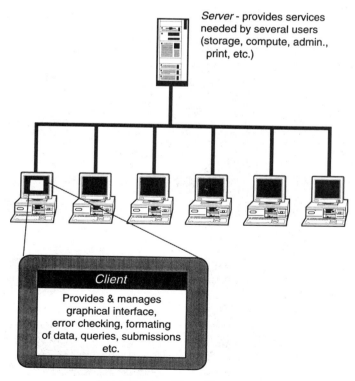

Figure 6-5 Client/Server system.

be requested by different criteria (field name, a field containing a particular piece of information, etc.). The functions in an application can be distributed as follows:

- *The Client:* This provides a graphical user-interface, making the system easier to use, and provides error checking to make sure that the structure of the request is valid and that the basis of the search is correct (if an error occurs, the request is rejected and a new query requested). When data is received, the client takes care of formatting the output (screen, printer, report, etc.).
- *The Server:* This performs the requested database operation and returns results. Minimal error checking is done since that function is provided by the client.

While this is a simple example, it illustrates the division of labor provided by the client and server. How the server accomplishes its role isn't important to the client, only that the message is sent and received. The location of the database and its implementation can change without disrupting the overall system.

Servers do provide other capabilities:

- *Administrative Servers:* These can provide a personnel directory, act as an electronic mail system manager, maintain password (including checking for expired passwords and the need to make changes), and so on. The key is that shared data is in one place, making it easier to maintain and provide protection against corruption.
- *Printing Servers:* These maintain print queues (lists of jobs to be processed), switching jobs between printers, etc.
- *File Servers:* discussed above.
- *Computer Server:* The controls access to high-performance computing and runs jobs.

Client/server (C/S) systems are fairly new to laboratories. Vendors have put a considerable amount of engineering into existing time-sharing-based systems, and the cost of redevelopment and uncertainties of the stability of underlying platforms makes reinvestment a point to be considered carefully.

The implementation of your computing strategy through the use of client/server systems can be an immediate one or an evolutionary one—working with a less sophisticated model until need builds and C/S becomes a need. Should you decide on the latter option, the software and hardware purchases you make should be done with that possibility in mind. Networks should be installed with a higher than immediately needed throughput to allow for growth and to avoid installing one system and then having to replace it later.

Applications software should be purchased with the potential to be used in a C/S environment.

INCREASING COMPUTING POWER

New features in operating systems (in particular, the move to graphical interfaces) and applications software put more and more of a load on the central processing unit (CPU) of the machine. Depending on what you are trying to do, there are several ways of improving the computing capability of a given machine.

Upgrading to a faster CPU is one alternative that generally works across the board an improving system response, but that isn't the only solution. If your work is math intensive, adding a floating-point unit (FPU—floating point arithmetic can be done either in software or hardware, the latter being much faster) can help. There are a number of cases where just putting in the additional hardware won't have any immediate effect; versions of software and math libraries have to be used that recognize the existence of the FPU.

The next step in increasing performance (once you've extracted all you can from one CPU) is to add another that is specialized for particular functions or a group of functions. In this realm are signal processing boards, array processors, and so on. In science applications, these have been common and cost anywhere from a few hundred dollars to several thousand. Most have gotten smaller over time and are now add-in boards.

In each case, the manufacturer has created a peripheral device that is passed data, does something, and passes back results. The device has an on-board CPU, memory, and ROM (read-only-memory) code that give fast calculations of FFTs for example. General array processors require programming, but accomplish the same thing. Add-in math processing boards also do the filtering needed for programs such as Adobe Photoshop (an image editing application). Problems such as contrast enhancement, image arithmetic, and so on have their performance improved by specialized hardware (Radius is one manufacturer). From the standpoint of computing strategies, parallel processing computers from Thinking Machines and supercomputers such as a Cray can be considered as a math engine for a desktop system ("my other computer is a Cray" was popular T-shirt for Macintosh users for awhile). These examples are common in computing. The philosophy is that if you need to get something done faster, get something that goes faster, there are other alternatives.

Distributing Processing

Solving problems requiring a lot of computing capability can be done by brute force (King Kong supposedly had the strength of 100 men—no sexism cant, that's what the script said). Another way of obtaining one-Kong's-worth of strength is to get 100 men; this does, however, present a management problem, getting them coordinated so that they can be effective in solving the problem. When we substitute "computer" for "men" in the previous sentences, we have the same problem. This section will look at some solutions for it. But before we get there, we have to solve one other issue—where do we get one-Kong's-worth of computers?

One computer may be sitting on your desk, another on your secretary's (assuming you have one), others are on the desk of your co-workers. You are using your machine during the day, but what is it doing at night: turned off or left on running the screen-saver and keeping the office warm? There is available computing capacity if it is made accessible to software that can make use of it. All it requires, aside from that software, is a network to allow the machines to communicate with each other and work cooperatively (sci-fi buffs minds are going into overdrive). How that cooperation takes place and what can be done is governed by the supervisory software (see Figure 6-6).

Distributed Batch Jobs

Consider the problem of having a large number of compute-bound jobs that have to be run on distributed systems. Each available system (desktop, workstation, etc.) is treated as a device capable of executing a specific type of computing problem. Each job is coded with its computing requirements (would it benefit from or require an array processor or other special hardware). The supervisory system has a list of outstanding jobs to be processed and the matching hardware requirements. Its task is to send jobs to the appropriate hardware on the network. When one job is done and its result

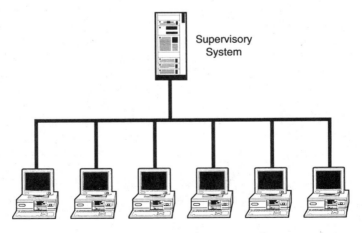

Desktop Systems - available during off-hours

Figure 6-6 *Supervisory system.*

INCREASING COMPUTING POWER

logged, the next job is sent. The supervisor is responsible for keeping the networked systems as busy as possible. The end result is that a large amount of computing can get done by parceling it out to otherwise idle computer systems.

Suppose you had a very large number of independent analyses to be performed. This would be a good way of handling that problem.

CPU Farms

Another approach to using available computing power is to take a single large job and break it down into individual modules that can be spread over a number of machines. Suppose a large array has to be work on, many thousands of elements. The array can be divided into a number of smaller problems, each processed by a different machine, and the result combined into a final solution.

The power of this approach is illustrated in the following excerpt from TIME Magazine (May 9th, 1994, page 19):

> **The Internet Factor**
>
> In a computational tour de force that could affect the security of the information superhighway, a team of computer scientists has solved a long-standing mathematical problem: finding the prime factors of a 129-digit composite number. When the puzzle was originally posed in 1977 by cryptographers trying to demonstrate the power of a new encryption system, scientists estimated it would take 40 quadrillion years to solve. But by using the Internet to divide the number crunching task among 1,600 computers, a team of volunteers managed to crack the code in just eight months. Corporations and government offices that rely on such codes may now have to shore up their systems for transmitting sensitive information.

If the problem can be divided into truly independent modules, supercomputing performance can be achieved on a collection of PCs and workstations. Information sharing between cooperating systems will degrade the performance, the amount being dependent on the amount of sharing and the characteristics of the supporting network—network speed and overhead are the key issues.

The major drawback to this approach is the work needed to demonstrate parallelism—the ability to divide the work into independent modules that can be executed without a significant dependence upon data from one another. One array example, which could represent a large number of measurements or a digitized photograph, is one in which a high degree of parallelism would be expected; improving the contrast or brightness of a photo allows us to treat each pixel as an independent element. If a calculation depends on knowing the value of a neighboring cell, then the effectiveness of this method is reduced. The appropriateness of this technology needs to be evaluated on a case by case basis.

Problems of broad interest, image enhancement for example, are benefiting from these techniques. The algorithms are well understood, there is a wide market, and vendors developing sharable systems to do the computations on otherwise unused computing power.

Much of the experience with this computing style comes from high-energy physics applications, where large number of data files need to be processed by the same program. Two papers published by Drs. Paul Avery and Andrew White give an example of the power of this technique (Paul Avery, "A New Approach to Distributed Computing in High Energy Physics", in *Proceedings of the XXVI International Conference on High Energy Physics*, Dallas, TX, Aug. 1992, AIP Conference Proceedings No. 272, James Sanford, Ed., pp. 1753–1757; and Paul Avery, "distributed HEP Computing using NetQueues", in *Proceedings of the 7th Meeting of the American Physical Society Division of Particles and Fields*, Chicago, IL, Nov. 1992, World Scientific, Carl Albright, Ed., pp. 1705–1707). The UFMULTI system consist of a series of VAX computers linked over Ethernet using DECnet as a communications protocol.

The task supervisor (see Figure 6-7) uses a list of available processors and a control file to manage a number of computers, each of which is running a

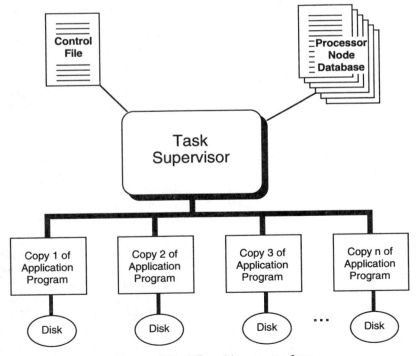

Figure 6-7 UF multicomputer farm.

INCREASING COMPUTING POWER

copy of the same applications software. Data sets are sent to each processor as it finishes its previous calculations. One experiment (CLEO from Cornell) had collected data from three million events. Each event consisted of 1000 bytes in packed format. Prior to using the multiprocessor system, it would take 24 hours to make a single pass through the data. Using a stack of 8 processors the analysis takes three and half hours. The estimated efficiency of the system is 85% (the loss is due to communications limitations).

Some of the processing power for the system came from desktop machines. During normal working hours, users of these systems did experience some delays in interactive use, so limitations were placed on the computing-intensive tasks (either suspended for periods of up to an hour or used during off-hours).

Linda Technology

An extension of this approach is referred to as "Linda Technology". Consider a situation where you have a number of objects that require processing. Each object is called a "tupple" and the collection of them is called a "tupple space" (see Figure 6-8).

The processing for each tupple is done by a worker who is specific to that type of tupple. A worker checks for the availability of its type of tupple object, does the processing required—which might create another type of tupple— and then goes back for more. This kind of problem occurs in database work, modeling, data analysis, and robotics: situations where you want to use a lot of computing capability one one type of application.

A distributed computer system, described above, can be used effectively in working this kind of problem. The tupple space would be available over a network, and each system (desktop PC or otherwise) would take on the characteristics of a tupple worker. Tupples would be requested and pro-

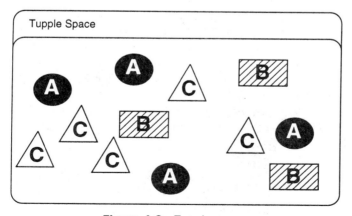

Figure 6-8 Tupple space.

cessed. The more workers (systems), the better performance you get. The supervisory system could keep track of the number of different types of tupples to be worked on and change the number of corresponding tupple workers (by downloading or executing new software) to maintain maximum throughput. As a result the system would be self-optimizing.

In the laboratory model described in the earlier chapters of this book, the combination of a Data Librarian and independent analysis modules makes the use of this type of technology attractive for processing large numbers of analysis. Data selection criteria applied against the Data Librarian define the tupple space, and processes for working—analyzing—the data are the tupple workers. These workers could reside on a single system or be spread over a number of cooperating machines, increasing the total number of workers. (*Note*: In a single shared processor this model is of little practical use since you are reduced to a single worker faced with a large number of tasks to do. Creating more workers on the single CPU just divides that CPU's time over more tasks and adds overhead processing, which reduces effective throughput.)

The concepts behind Linda technology are common in human experience. The working of a laboratory (a testing laboratory in particular) is one example. The "tupple space" is the work to be done, the laboratory professionals, the worker. It is a self-optimizing system since people take different roles depending on the work load. Management of the lab mimics the tupple supervisor, adding "workers" as needed or improving their ability to do the processing, optimizing throughput as a function of cost. In a fully automated quality control laboratory, as noted in the section on sample management and robotics, the samples are the tupple space and robots and automated instruments the workers.

Tupple spaces and tupple workers are the technology of the assembly line and automated factory. While we usually don't apply those principles to laboratory work, in some instances—large volume routine testing for example—they are appropriate. That approach would streamline lab procedures, reduce the need for people-as-machines, and allow them to spend their time doing work that is more appropriate to a human intelligence, work that is challenging. If you seriously consider that direction, also consider the needs, feelings, and future role of the people who are currently doing the lab work. Machines replacing people has never been a popular theme, people *supplemented and aided* by machines could be.

The performance of these systems is dependent upon the individual application: how well it can be parallelized and the effectiveness of the communications medium.

HYPER-INFORMATION SYSTEMS

Hyper-information is not a new category of knowledge/information/data but refers to the ability to work with K.I.D. in new ways. The classic paper on concept of hyper-information (cited by almost every publication on this subject) was authored by Vannevar Bush ("As We May Think", *The Atlantic Monthly*, July 1945, No. 176, pp. 101–108) in which he describes a machine called MEMEX. MEMEX was to act as a repository for information and would allow cross-referencing and access to material. An article that referenced other material within this system would have that secondary material readily available. A concept or phrase that needed more explanation would have that associated data made available to the reader. *Star Trek's* ship-wide computer had the same capabilities (with voice and video I/O).

There have been a number of efforts at putting that technology into at least test-bed use. Brown University's efforts resulted in the IRIS project (Institute for Research in Information and Scholarship). A review of that work is published in IRIS Technical Report 87-4 ("Designing Hypermedia 'Ideabases'—The Intermedia Experience," Nicole Yankelovich, George Landow, Peter Heywood). The January 1988 issue of IEEE Computer (vol. 21, No. 1) contains a second reference to that work ("Intermedia: The Concept and the Construction of a Seamless Information Environment", Nicole Yankelovich, Bernard J. Haan, Norman K. Meyrowitz, and Steven M. Drucker, pp. 81–96) as well as the description of another system ("Finding Facts vs. Browsing Knowledge in Hypertext Systems", Gary Marchionini and Ben Shneiderman, pp. 70–80).

Some steps were taken to provide commercial software with the ability to create software systems capable of pointing and linking structures. The most notable is Apple Computer's HyperCard, followed by a number of similar products on Macintoshes and PCs. Harvard Graphics presentation software had some linking capability, letting you move through a slide presentation based on areas of interest rather than a linear sequence. Kodak's Photo CD system has software that allows the user to create presentations with branch points, allowing the user to access different portions of a presentation depending upon your interest. The Grolier CD-ROM Encyclopedia lets you build notebooks of information structures depending on the subject you are studying. Most of the software built on the products noted consists of demonstration systems and help files using the linking and graphical interface as a substitute for a menu, rather than the system V. Bush envisioned. A real hyper-text system would be context dependent, and some examples do exist. Voyager Systems has been producing bounded hyper-information systems (information access is limited to the data sets provided). The most recent system is an interactive version of Macbeth including video, sound, text, links to related material, and a "hypertext" feature that allows you to click on a word and get its pronunciation and definition. True general-purpose hyper-information systems have yet to be developed; they would require a degree of standardization for data, text, video, sound, etc., that doesn't exist yet. Bounded commercially available systems on CD-ROM avoid that issue by encoding all the data internally.

Why is this important to you? Better access to, and control over, information and data has been on the wish list of most organizations ("How

Information Gives You Competitive Advantage", Michael E. Porter and Victor E. Miller, *Harvard Business Review*, July–August 1985, pp. 149–160). "Information glut", a complaint of many companies, refers not to the problem of having too much, but to having it in a form that is unworkable, incompatible with other information forms, and difficult to synthesize and use. Libraries are full of the stuff, and card catalogs are overwhelmed by the arrival of material. When was the last time you remembered seeing "something" on a topic and then tried to locate it? How many stacks of periodicals did you go through? Did you find it? Did it survive the last attempt to clean your office (*it used to be over there...*)? While the potential of hyper-information systems hasn't born fruit yet, it eventually will.

Laboratories are no different than any other organization when it comes to the need to improve information access and management. To give you an idea of how hyper-information technology might be put to use in a lab, let's recast an example from the IEEE Intermedia article (illustration p. 83) in a more appropriate setting.

You need to do an analysis for *Agent X*, an antioxidant in low-density polyethylene. After asking everyone else first, you decide to do a literature search. You enter the subject of the search into the system and look at the options (see Figure 6-9).

Clicking on any option, brings you the appropriate information. Database systems do similar things today. What hyper-information potentially offers you is the ability to link information into networks that are meaningful to you. One object can point to another, so that retracing your way through the material is quicker. That network of linkages persists after the session is completed, so that the next time you want to check on that topic the relationships you determined are still in place, ready to be used as is, or extended, perhaps

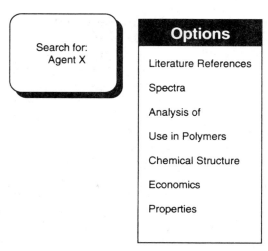

Figure 6-9 System search.

HYPER-INFORMATION SYSTEMS

through the analysis of a related material. The value of the information increases because of the relationships you have developed. Now, rather than dealing with a collection of reference information, you are looking at information with a higher level of meaning, hence hyper-information. Information from commercial sources can be linked with data developed internally.

This networking structure would be particularly valuable in a laboratory notebook. Your work could be cross-referenced within itself *and* referenced to the work of others (see Figure 6-10). The notebook becomes a collection of your data plus the linkages to other notebooks (security and privacy issues permitting) and can inherit the linkages they have made.

Even though we are talking about future systems, those "futures" do have significance for us in the determination of a strategy. Suppose an electronic lab notebook with hyper-information capability were announced for availability in a year. Is your laboratory data in a form that can take advantage of it? Even if the format isn't known, you can take some steps.

First, you can make sure that all relevant data is maintained in electronic form, in a standardized data format. All text, for example, should conform to a particular word processor scheme, spreadsheets should be built around a common application, and data from instruments should be in a standardized data format. Second, files should be organized and archived on a server for ready access.

With those two steps, you have put in place a strategy for maintaining and

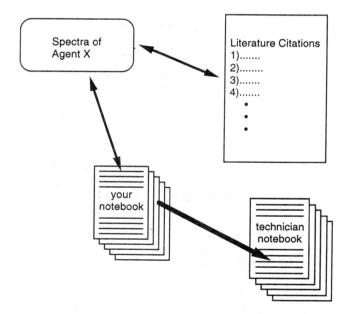

Figure 6-10 Networking structure for a laboratory notebook.

managing data. When such a product comes about, the conversion from whatever format you are using to whatever format the vendor will require will be based on a small number of data formats. Richard Weimar (at the 1991 Scientific Computing & Automation Symposium) in one presentation noted that an informal survey of a large chemical company showed that researchers can access over 100 databases, both corporate and individual. How are you, or your company, going to manage the number of database system conversions to take advantage of new software technology? The likely answer is you won't. New software, like an electronic lab notebook, will be phased-in, new data will be entered into the notebook, old data will exist as is, and it will take years before the phasing is completed. While standardization may be considered a straightjacket by some, limiting options, considered standardization in fact provides the basis of flexibility.

There are implications for storage systems as well. Suppose we had a general-purpose hyper-information facility today: one that would link memos, mail, your calendar, data, graphics, presentations, etc. Your 10:00 am meeting would be linked to the memo containing the agenda (click on the icon and up pops the text). Related memos would be linked to replies and counterarguments. All the material for your next major presentation is linked in a network of linkages that allows you to get at anything you ever wanted to know. If you drew a map of all the linkages in this system, it might look like a pile of spaghetti.

In order for this system to continue to function, all of that material needs to be available to programs that need it. You can't move anything off-line (on tapes or disks that need to be manually mounted/dismounted) without disrupting the network. If that network is limited to your personal information, you can judge for yourself the ramifications of that happening. If it is shared information, then there may not be any way of determining the extent of that disruption. Moving data to CD-ROMs or write-once optical storage is possible, but the supporting software would have to maintain a separate database of linkages on high availability read/write media (you can't write the linkages on to these media). The entire linkage database may be designed as a separate data structure for that purpose, making it easier to deal with through automated mass-storage systems. In that case, the data structure would have to be put on very fast media to avoid performance degradation of the system. In a shared network environment, the performance of the network has to be tuned to avoid having it become a bottleneck.

If you consider the value that the linkage database represents (in terms of data access and the cost of recreating it), that fast medium should have a shadow disk (or whatever storage technology is used) so that a duplicate copy is maintained and updated on every change. This would protect you against the loss of the linkage database if the primary medium failed.

Technologies like this are only valuable to you as long as they are functional. That includes the underlying software and the assurance that all the underlying and supporting hardware and software systems are protected.

Uninterruptable power supplies, redundant disks, automated backup, reliable network support, and backup systems are all part of that requirement. Of greater importance is the need for carefully considered planning, strategy development, and cooperation between laboratory and information services groups. The cost of these technologies (hardware, software, and people) is too large for a typical laboratory to support by itself. The benefit of these technologies extends well beyond the laboratory, as does the need for data sharing. Cooperative planning between groups is mandatory.

NETWORKS AND COMMUNICATIONS

Throughout this book reference has been made to the need for effective communications between systems; in automated systems, that communications means networks of computers and equipment. In modern laboratory automation, in all but the smallest laboratories, networks substantially improve the ability for people to work.

If this book were only about office systems, including laboratory offices, the material would be quite short; networks are good—determine the speed and capacity you need, the products you want to work with and off you go. As noted earlier, bookstores are full of texts on computer communications, networks, and so on. We can afford that short a summary because in the case of computer-to-computer communications there are good reliable products on the market that take a once-upon-a-time demanding problem and replace it with a plug-and-play system. These systems can be just that easy to work with, the simplest is the Macintosh.

Each Macintosh comes with networking software built in. It's accessible through the printer port over a serial link or via Ethernet using separate hardware. In both cases, the Appletalk protocol is the same. Setup takes a few minutes and with little effort you get

- *Network Access to Printers:* Appletalk compatible printers, including laser, ink-jet, and dot-matrix are available to several users.
- *File Copying:* Files can be transferred between machines as easily as between folders on one machine; the drag-and-drop interface is simple to use.
- *Shared Access to Devices:* Hard disks and CD-ROM are available as shared devices.
- *Support:* For protected access to files, folders, and volumes.

There are a few flaws (a mount/unmount function would be nice) but it is easy to use and set up. Laboratory systems are different.

Making Connections

One major difficulty comes from the reliance of instrument vendors on serial communications through RS-232, RS-422, and RS-485 or the higher capability IEEE-488 (and its follow-ons) communication protocols. They almost imply network capability, but in practical use fall short. Serial communications is used successfully in networks—the Macintosh uses it. The difference is that in the true networks, protocols exist for message transfer, error detection and correction, etc. Those standardized capabilities are lacking in most instrumentation.

In the mid-1970s there was no single strong contender for a dominant operating system or hardware vendor. Networking was in its infancy. The primary method of getting measurements into a computer was via the analog-to-digital converter (A/D). Competitive pressures and market requirements gave the edge in instrument sales to vendors who could provide "computer compatible" devices (eliminating the need for an A/D). There were two choices: parallel transfer via digital I/O cards or serial I/O. Parallel I/O, available on many devices, was expensive to use and usually limited to short distances. It also required the purchase of boards dedicated to that purpose.

Serial I/O (there were two versions: 20 milliamp current loop and RS-232) cost less, used less expensive wiring, and used the same type of interface used to connect terminals to computers. Programming for the serial port was well understood, and "computer compatible" came to mean serial ports. Some terminals came with a pass-through feature (allowing characters from one device to share the same line as the terminal—in that case, programming the device was the same as programming the terminal) or used a manual switch to choose between the instrument and terminal.

With serial I/O as the choice of communications method, physically connecting computers to an instrument was an exercise in understanding the users manual (many vendors varied from the RS-232 specification), adding the needed programming, and using the device. Every instrument and computer connection was a custom effort.

The situation today isn't much different—although the NIST/CAALS effort is one opportunity to change the situation. Instruments in general, some have built in workstation/network support, are not full direct participants in laboratory networks. Their data flow must be directed through an intermediate system acting as a controller and network interface/protocol converter.

That device can be a PC used to control the device, acquire data, reformatting it if necessary and transmit it to another system. It can also be an interface box like Digital Equipment's Corps. Local Area Transport (LAT) connection box (see Figure 6-11). The original purpose of the LAT was to provide a means of connecting terminals to networked computers over Ethernet and avoid the need to string individual sets of wires to a particular

NETWORKS AND COMMUNICATIONS

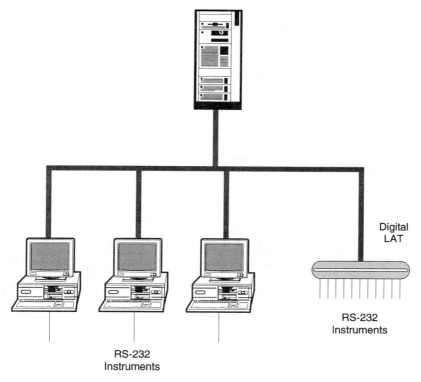

Figure 6-11 LAT connection box.

computer or common switch system. As new versions of the product became available, new capabilities were added. At one point a critical barrier was passed: a host computer could initiate a connection to a particular port on a particular LAT without having to share that port with a terminal (the terminal had been needed in earlier versions to initiate the connection). That meant that any computer on the network can talk directly to any serial device including devices that converted RS-232 character streams to IEEE-488 controllers (usually only one computer would work with a particular instrument to avoid confusion, although one computer could connect to several instruments simultaneously).

That did, and still does, have far-reaching implications for laboratory instrumentation:

- *Remove Access:* Lab personnel don't have to be in the same room or the same building to check on the status of an instrument or experiment.
- *Centralized Systems:* Systems can be designed to handle most laboratory computing requirements—you don't need a separate $2000 computer to service a $600 device (you will need some device to let the host

computer know that the instrument is ready for work, a simple terminal will do).

While these benefits look attractive, they have to be balanced against risk. If everything is loaded on one machine, what happens to your lab if the machine or the supporting network fails? Putting a number of instruments on the network can significantly increase the number of message packets transmitted and degrade the network's performance.

While networking has become easier to manage and design, the best results come from careful planning and not being afraid to spend dollars to get the best solution; an inexpensive poor solution is paid for on a daily basis.

Network Design and Management Considerations

The development or evolution (planned vs. "just happens") of the network will pass through several stages. The success of these transitions depends on the planning put into the system. At first the network becomes an easy way to pass files and messages. Each computer acts as an independent entity with only casual acknowledgment that it is part of a community. At this stage, disk crashes or new software being added to a particular machine does not have any broad impact—though considerable care should be taken to prevent viruses and their kin from affecting the system.

At the next step, commonly used files are stored on a common disk, either on one of the desktop systems or on a separate server. Backup is done, occasionally. The server use grows until someone realizes that a great deal of valuable information is stored on it, and that its loss will significantly affect the operation of the laboratory. If the laboratory network is independent of the rest of the company, someone will take on the responsibility of system manager—less science, more computers (this is how a lot of scientists became programmers).

Finally people realize that network transactions are a major part of the lab's operations. The laboratory, using LIMS or other software, has now become dependent on the network for day to day results. At this point a couple things probably have happened. First, you either have a full time lab person acting as system manager or you have developed a relationship with the MIS group. Second, the systems on the desktop are no longer isolated computers, but are part of the overall system. Each is capable of running software, and each has begun to be viewed as a port into the network laboratory system; it is no longer "my" desktop system it is part of *the* system. The extent of the network and the traffic on it help define the work flow. Disruptions to the system have catastrophic effects on the lab's ability to function and meet its goals.

The point of that description is this: laboratories will become dependent on networks, and the success of those networks will be dependent on proper management. "The network is the system" is a marketing catch-phrase that has been used by a number of vendors. It is true. The computer on your desk

NETWORKS AND COMMUNICATIONS

is part of the company network. Access to resources (printers, etc.) is governed by access to the net, and a failure of a network component effects everyone. Do you want your people to become systems managers, or do you want to have the computers managed by a corporate group? That is a question you will need to consider early on. Who controls the network and what functions/features will be accessible to you?

Throughout this book we've looked at the impact of communications on lab systems. Network design and management is where you decide the limitations on those systems. Once systems are in place, changes will be painful and expensive.

Network Considerations

Some of the points we are going to cover have been touched on earlier, but this is a good place to sum things up.

The *backbone* (see Figure 6-12) is the major interconnect of the entire system. The double-headed arrow is there to signify that it extends beyond the laboratory. The width of the line is what it is for two reasons: it has to have a much higher carrying capacity than any of the attached systems, and the word "backbone" had to fit into it (the capacity issue being the key one). Just as the interstate highway system roadways are larger and faster than their feeder roads, the backbone has to carry the combined load of all its feeders or else it becomes a bottleneck (just as a traffic mishap can backlog a highway). When you are initially designing a system, you need to plan for worst case situations. The backbone should be many times the capacity of the feeding systems to handle their load and allow for growth. You may also want to consider a duplicate backbone to off-load peak traffic and serve as an alternate backbone if anything happens to the primary (the New Jersey Turnpike's dual roadways are constructed that way, in part, for a similar purpose. The analogies with automotive traffic flow are excellent, so if you want an idea of what can go wrong with a network, take a drive on the interstate at rush hour).

System 1 in Figure 6-12 is a semi-autonomous workstation. It is a fully functional, self-contained system, but is must rely on the network for access to print services. The individual running this system must be able to manage the system, install his own software, provide backup, etc. The manager of System 1 is also responsible for ensuring that any new software will be virus free and that it will not adversely affect the performance of the network. All the software needed, aside from the LIMS, is available from the system's disk, so it is immune from most network-wide problems. This is the type of system most people today work with, but that may change.

Systems 2, 3, and 4 are full systems consisting of displays, keyboards, mouse, disks, and memory. The disk on each system is smaller than that on System 1. Disk space is used for local data storage and minimal system overhead, operating system and applications software are stored and accessed from the local server. This approach has some gains and losses over the

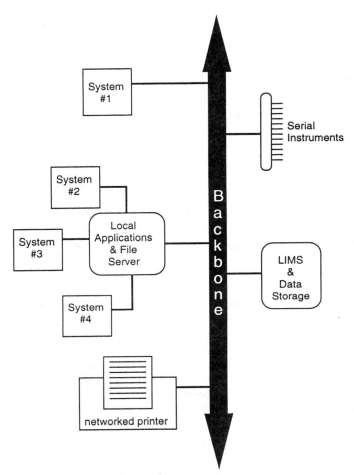

Figure 6-12 Network design.

design of System 1; their impact needs to be measured against your goals—some people would look at some of the "losses" as a problem, others would be relieved. The people using these systems are not responsible for system maintenance, for installing software, upgrading to new versions, virus, protection, and most backup functions (copies of data files on their local disk need to be backed up; that can be automated if necessary). That can be viewed either as a gain or loss depending on each person's viewpoint. On the other hand they are dependent on someone else to maintain and manage the systems. Problem resolution may take longer, and systematic problems can affect all users of that system/server combination. If there is a network problem or a problem with the server itself, all the connected systems can be rendered unusable until those issues are fixed. How do you want people to

NETWORKS AND COMMUNICATIONS

spend their time? How do you trade-off the risks against the benefits, and how do you protect yourself from problems? Some companies will rely on the MIS department to manage systems in this configuration. That will mean competing for resources when problems arise. It may also mean a lack of flexibility on hardware and software choices. In order to reduce the maintenance costs, MIS departments may choose to support only certain hardware systems and software packages (this may make overall systems validation easier). Restricting choices may not sit well with the users, particularly in laboratories. Data system vendors may recommend or require that computers be purchased from a particular set of vendors and that may conflict with MIS's choices.

Vendor specifications need to be taken seriously. In at least one case, a vendor is specifying particular computer vendors. This is not due to marketing relationships, but because the vendor has determined that its software will work on some machines and not others. Varian has determined that some PC ROM software (specifically the BIOS) on some PC systems will not work with its software. As a result, Varian cannot guarantee system integrity (this was reported by a Varian representative at the LASF's September 1993 Instrument and Data System Validation Symposium). If you deviate from the vendor's specifications, you may not be able to adequately validate the system—you can be sure that the inspector will ask why you ignored the vendor's recommendation. MIS groups need to understand that vendor selection criteria need to be expanded to solve laboratory problems.

Figure 6-12 shows *LIMS and instrument access* being gained through the network. Any disruption of the network will have serious consequences for these functions. That leaves you with two basic choices: don't limit access to one method or take pains to ensure that the system will be able to survive most problems. In the case of instruments being serviced over a network, this is not a trivial concern.

When everything is working well, the only concerns are making sure they stay that way. Should the network fail, the problems are more than just not being able to run the systems, there is the issue of recovering from failures—the most common of which are intermittent problems. Suppose we have a system that is running smoothly and then a problem occurs: maybe the system hangs, a power spike causes a loss of communications, etc. What are the ramifications of that loss? The instruments may continue to chug along merrily—the temporary loss of communications may not have been noticed—*but* you will have lost synchronization between the controlling software and the experiments. Have you built in fail-safe mechanisms into the instrument-level software? Can the instruments gracefully put themselves into a safe mode automatically if a message isn't responded to in a given time frame, or is manual intervention required? Does the data lose its validity? How do you regain data synchronization between the instruments, data systems, and LIMS? How robust is the system design?

Laboratory network design and management is not simply a matter drawing a connection diagram and choosing products. It is a matter of managing a dynamic real-time environment and planning for the consequences of failures. Office systems don't face the same issues nor do failures have the same potential consequences. Again, it is an area where the cooperation of MIS group is needed, unless you are going to fund your own independent system. The MIS group needs to understand that laboratory systems are not just a different flavor of office systems. Shutting down the network at night to handle maintenance problems may have a disastrous effect on an automated lab running 3 shifts or designed for unattended operation. Software needs to be designed to handle worst case situations and not assume that things will always run well. The fact that something is technically possible doesn't mean it's the right choice.

APPENDIX

Analytical Instrument Association Chromatography Data Standard Specification

Developed by the
Analytical Instrument Association
Committee on Communications Standards

PREFACE

Located in Alexandria, Virginia, USA, the Analytical Instrument Association (AIA) is a trade association for manufacturers and suppliers of analytical instruments and related products and services used for chemical and biomolecular analysis. AIA members' sales currently account for 60% of the global market and 80% of the U.S. market for analytical instruments. The goal of AIA's standards development work is to benefit users by helping to standardize data interchange and storage for analytical instruments and laboratory software across different vendor's products and platforms. The AIA Communication Standards Committee is comprised of representatives from the AIA member companies working together and supporting the philosophy of open systems for the analytical instrument industry, which is already prevalent throughout the computer industry.

The overall objectives of the standards development work are:

1. develop robust, workable specifications in a timely fashion
2. ensure simplicity and measurable success for the first versions
3. ensure the approach and specifications extend to other analytical techniques

The first focus for the AIA efforts is the family of chromatographic techniques.

As the AIA Committee on Communication Standards does more research on the needs of the industry and problems of general data portability, future versions will be implemented with more robust data encoding and access methods, that accommodate more types of information, including information from other analytical techniques.

This Specification is one of two AIA documents in the AIA Chromatography Data Standard distribution kit. 'Me AIA documents include:

1. **AIA Chromatography Data Standard–Specification**–Gives the full definitions of the data elements used in implementation of the AIA Chromatography Data Standard.

2. **AIA Chromatography Data Standard–Implementation Guide**– Gives the full details on how to implement the content of the AIA Chromatography Data Specification using the public-domain netCDF data interchange system. It includes a brief introduction to using netCDF. It is intended for software implementors, not those wanting to understand the definitions of data in a chromatography dataset.

This Specification gives complete definitions for each one of the generic chromatography data elements. It defines the Analytical Information Categories, which are a convenience for sorting analytical data elements to make them easier to standardize. This document is intended to be used by anyone wanting to fully understand the definitions of the AIA Chromatography Data Standard.

ACKNOWLEDGMENTS

The Analytical Instruments Association, in the form of its Board of Directors and member companies, undertook a unique challenge starting almost three years ago: the formation of a special committee to investigate and then create a computer-to-computer communications specification for exchange of data among analytical instrument systems and other data gathering and manipulating computer systems.

It was not at all clear that this activity would be successful, or even desirable. As the committee met and deliberated, its progress was followed by frequent reviews by the Board of Directors and reports to the membership.

What began to emerge was a pattern of dogged determination and grass roots work by the committee technical representatives, backed by unflagging support from their management... in spite of a flat instrumentation market during most of the time of the work.

ACKNOWLEDGMENTS

It is clear now that we will publish a specification that can be used as part of a common data interchange protocol (because it is being used now by member companies who are testing it). It also is clear that the enthusiasm for this work by the contributing workers and by others (users, other vendors) is greater than ever.

The Analytical Chemistry community and the membership of the AIA is indebted to many people who have contributed to this work, directly or indirectly.

We acknowledge here the working members of the committee and their companies. We apologize for any people not listed who may have contributed along the way.

ANALYTICAL INSTRUMENT ASSOC.	Mike Duff
BECKMAN INSTRUMENTS	John Lillig
	Ted Lo
	Ian McCutcheon
DIGITAL EQUIPMENT CORP.	Rich Lysakowski
	Paul Strauss
DIONEX CORP.	Bart Evans
	Der-Min Fan
	Sue Strong
HEWLETT-PACKARD CORP.	Vince Dauciunas
	Dan Holmes
HITACHI INSTRUMENTS INC.	Tom Zarelia
	Bob Kuhar
	Jeff Bernhardt
PERKIN-ELMER CORP.	Mike McConnell
	Dave Nelson
SHIMADZU SCIENTIFIC INSTR.	Bob Patch
	T. Nishimura
	Tim Gizinski
SPECTRA-PHYSICS ANALYTICAL	Barry Tomlinson
VARIAN ASSOCIATES	Jerry Keefe
	Jean-Louis Excoffier
WATERS CHROMATOGRAPHY	Rohit Khanna
VG/FISONS	Mike Head
	Kevin Smith

In particular, Rich Lysakowski deserves special mention because of his leading technical role in carrying the work forward, for evaluating and suggesting netCDF as the vehicle for data interchange, and for his objectivity in compromising his long term ideals, which are forward looking and important, in favor of the shorter-term need of the Committee to produce a working first specification.

The group is indebted to Rich Lysakowski for originating this document and the others, and for their revisions as changes were required. Final editing was completed by Mike Duff, Vince Dauciunas, Dave Nelson, Rich Lysakowski, Der Min Fan, and Mike McConnell.

Also, we would like to single out Bart Evans and Der-Min Fan for their continued aggressiveness in pursuit of the standard... they hustled the group along and added much to the need for action and progress. Der Min also made extraordinary technical contributions in helping select and define data elements for the specification.

Mike McConnell has been especially helpful in interpreting chromatography needs and picking through the labyrinth of logic related to the actual data elements, their meaning, and their placement in the information categories.

Vince Dauciunas is to be acknowledged for keeping the Technical Committee meetings logistically focused, for his expert ears in capturing the collective wisdom of the group during the design sessions, and for his own technical contributions.

Finally, the work of the committee could not have proceeded without the constant support of the Board of Directors of the AIA, who could have backed away from the effort at any time were they less willing to take the risk that something useful would actually get done.

David Nelson
Chairman,
AIA Committee on
Communications Standards

TABLE OF CONTENTS

Preface 177

Acknowledgments 178

1 Introduction 183

2 Analytical Information Categories 186

3 AIA Chromatography Data Standard
  Specification V1.0. 188

Appendix A–References 202

List of Figures

Figure 1 – Raw Data Element Semantics 196

List of Tables

ADMINISTRATIVE Information Class 188
SAMPLE-DESCRIPTION Information Class 193
DETECTION-METHOD Information Class 194
RAW-DATA Information Class 195
PEAK-PROCESSING-RESULTS Information Class 198

INTRODUCTION

1.1 AIA Data Communication Standards Development Work

The Analytical Instrument Association's Committee on Communication Standards is producing specifications, a software implementation, and documentation for a generic system for analytical data interchange and storage. A brief discussion of the technical objectives is given to place this work in proper perspective.

1.2 Technical Objectives of the AIA Chromatography Data Standards Work

1.2.1 STANDARDS DEVELOPMENT AND SYSTEM SELECTION

The technical goals of the AIA Committee have been to develop a standard for analytical data representation and interchange that meets the following criteria:

1. easy to use by software developers and end users
2. readable by humans using some facile mechanism
3. open, extensible, and maintainable
4. applies to multidimensional data (for hyphenated techniques) as well as two-dimensional data
5. independent of any particular communication link, i.e., RS-232, IEEE-488, Local Area Networks, etc.
6. independent of a particular operating system, i.e., DOS, OS/2, UNI@ VMS, MVS, etc.
7. independent of any particular vendor, and acceptable and usable by all
8. coexists with, and doesn't negate, other standards
9. designed for the long-term and implemented for use in the short-term, i.e., "Walk before run"
10. works well for chromatography and does not preclude extensions to other analytical technique families.

1.2.2 DATA INTEGRITY ACROSS HETEROGENEOUS SYSTEMS

The current implementation specifies a mechanism with particular directionality for data transfer integrity. The AIA Chromatography Data Standard has unidirectional data integrity for data transfers between heterogeneous systems. This is because source systems and target systems are made by different manufacturers, or if the systems are from the same manufacturer, they may use different hardware or algorithms. An example would be data transfer from Vendor A's data system running on a DOS-based personal computer to Vendor Z's LIMS running on a Unix-based minicomputer; another would be transfer between chromatography data systems made by different manufacturers.

If the receiving system has algorithms that assume a different analog-to-digital (ADC) converter word length from the sending system, and it calculates results based on its own, different data precision and accuracy, then the accuracy and precision of the original data is not going to be maintained. For example, if the sending system has an algorithm that assumes a 24-bit internal representation, and the receiving system has an algorithm that assumes a 20-bit internal representation, one may lose data accuracy and precision. If calculations are done by the receiving system, and the data are then sent back to the source system for further calculations, data integrity may not be maintained. Thus, there is an inherent directionality to data transfer given by different algorithms and different hardware systems.

The AIA Chromatography Data Standard can be used for data round-trips relative to the source system, e.g., from the source system to an archive and then back to the source system again. Such round-trip data transfers will maintain data integrity as long as there was no calculation or alteration of the data during transfer that would alter its accuracy or precision.

Thus, the AIA Chromatography Data Standard is bi-directional for homogeneous source-system round trips and unidirectional for heterogeneous source-to-target transfers.

The first implementation allows transfer of chromatographic raw data and final results. Plotting and requantitation of raw data on other vendor's data system (for comparison purposes); and transfer of final results to information systems (such as a LIMS) are possible in the first implementation.

1.2.3 ALGORITHMIC ISSUES

Algorithmic issues are not addressed at all by this Specification. Users cannot expect to get the same exact processed results from systems that use completely different algorithms.

1.2.4 ABSOLUTE SCALING OF RAW DATA

Absolute scaling of raw data across different manufacturers' systems is not possible at this time, due to the lack of general-purpose algorithms that can convert and scale data of different internal representations, from different data acquisition systems, and different computer hardware systems.

INTRODUCTION

1.2.5 GLP AND ISO 9000 REQUIREMENTS

AIA Chromatography Data Standard does not yet specify all elements needed to meet documentation quality data requirements (e.g., Good Laboratory Practices or ISO 9000).

1.3 Technical Features of the AIA Specifications

1.3.1 SEPARATION OF CONCEPT FROM IMPLEMENTATION

There is a clean separation of AIA Data Standard into *contents* (the data definitions within a data model) and container (the data interchange system). This is important because it effectively decouples concept from implementation. Computer technology is changing much more rapidly than analytical data definitions, which are stabilizing for the maturing analytical instrument industry. Producing an Accurate analytical information model and having well-defined definitions for data elements within that model actually have higher long-term significance than any particular data interchange system technology.

1.3.2 GENERAL TECHNICAL FEATURES

Two general technical features of the AIA Committee's Standards program stand out:

1) The Analytical Information Categories—a convenience for simplifying the work of developing analytical data specifications. These five categories were chosen based on three practical considerations: 1) which data is of interest to transfer most routinely, 2) which data can be standardized most easily in the short-term, and 3) which data can be standardized in the long-term. The analytical information categories are explained later in this document.

2) The Data Interchange System—the container used to communicate data between applications, in a way that is independent of both computer platforms and end-user applications. The system has software routines that are used to read, write, and manipulate data in analytical datasets. It has a data access interface, called an Application Programming Interface (API).

The data interchange system that most closely fits the scientific and software engineering requirements for a public-domain data interchange software system is the netCDF (network Common Data Form) system. The Unidata Corporation, which supports the National Center for Atmospheric Research, is the source of netCDF. netCDF is copyrighted by the Unidata Corporation. The AIA is using the netCDF system for the implementation of its Standard. The engineering 'beta" tests that prove applicability of netCDF for analytical data applications have been completed. An overview of netCDF is given in the companion document *AIA Chromatography Data Standard Implementation Guide (1)*. For more detail on the netCDF system, consult *The netCDF User's Guide (2)*.

2 ANALYTICAL INFORMATION CATEGORIES

Early in this work it was recognized that data and information usage varies widely in complexity and completeness. Information was sorted into logical categories, called the "Analytical Information Categories." These categories serve two very useful purposes.

First, the categories sort analytical information into convenient sets to allow more rapid standardization. This has made it easier for implementors to produce working demonstrations, without the burden and complexity of the hundreds of data elements contained in a full dataset for any given analytical technique.

Second, the categories accommodate different organizations' usage of information more easily. Some organizations may want only to transfer raw data among data systems. Others may want to transfer information to a LIMS or other database systems. Still others may want to build databases of chemical methods, instrument methods, or data processing methods. This first version of the Chromatography Data Standards is for a single sample injection, not for sequences of samples.

The information contained in this Specification represents the greatest common subset of information end-user and vendor requirements available at this time.

The Supplement sets forth the work-to-date done by the AIA Committee to define Categories 3, 4, and 5. The Supplement is included so the user can foresee the extension of Categories 1 and 2 to the full data-set.

2.1 Category I—Raw Data Only

Category 1 is used for transferring raw data. It includes raw data, units, and relative data scaling information. This will allow accurate replotting of the chromatogram and/or reprocessing.

Category 1 also contains administrative information needed to locate the original chemical and data processing methods used with this dataset.

ANALYTICAL INFORMATION CATEGORIES

2.2 Category 2—Final Results

All post-quantitation calculated results are included. This information category includes the amounts and identities (if determinable) of each component in a sample. Final sample peak processing results, component identities, sample component amounts, and other derived quantities of interest to the analyst are included in Category 2 datasets. Quantitation decisions are included here *as comments* to aid the analyst in determining how the results were calculated.

Category 2 datasets can be used to transfer data to database management systems, such as a LIMS, research databases, or sample tracking systems. It can also be used to transfer data to data analysis packages, spreadsheets, visualization packages, or other software packages.

2.3 Category 3—Full Data Processing Method

Quantitation decisions and data processing methods are transferred in this category. Quantitatively correct data transfer is achieved by this category for all parameters necessary to do peak detection, measurement, and response factor calculation, and calibration for a *sequence* of related sample runs. This applies to both samples and reference standards. Sample quantitation results are not included here; those are in Category 2. Peak processing method parameters, response factor calculation and other calibration method parameters required to quantitate sample component peaks are included in Category 3.

2.4 Category 4—Full Chemical Method

All chemical method information needed to repeat the experiment *under exactly the same chemical conditions* is included in this category.

2.5 Category 5—Good Laboratory Practice Information

Any additional information required to satisfy Good Laboratory Practices or ISO-9000 requirements are included in this category. This category generally deals with capturing product, process, and documentation quality information needed for validation.

At publication time, the AIA Committee has implemented the full contents of Categories 1 and 2 plus some additional data elements (mostly Administrative and Sample Description Information) from Categories 3, 4, and 5. You will find the implemented data *elements in Section 3 of this document.*

The AIA still needs to decide whether it is more appropriate to standardize information Categories 3, 4, and 5, or to create Categories 1 and 2 for other analytical techniques. These higher categories are generally useful for building and using databases of chemical conditions, archival of full experiment and test data, requantitaling analytical data, and other data processing information, etc. More extensive feedback from end users and additional vendors is being sought to help resolve these questions.

3 AIA CHROMATOGRAPHY DATA STANDARD SPECIFICATION VI.0

This section contains the definitions for those data elements which have been implemented by the AIA members.

Notes About Data Elements

Particular Analytical Information Categories (Cl, C2, C3, C4, or C5) are assigned to each data element under the *Category* column. The meaning of this category assignment was explained in Section 2.

3.1 Administrative Information Class

ADMINISTRATIVE INFORMATION CLASS

	Data Element Name	Datatype	Category	Required
1.	dataset-completeness	string	C1	M12345
2.	aia-template-revision	string	C1	M12345
3.	netcdf-revision	string	C1	M12345
4.	languages	string	C5	
5.	administrative-comments	string	C1 or C2	
6.	dataset-origin	string	C1	M5
7.	dataset-owner	string	C1	
8.	dataset-date-time-stamp	string	C1	
9.	injection-date-time-stamp	string	C1	M12345
10.	experiment-title	string	C1	
11.	operator-name	string	C1	M5
12.	separation-experiment-type	string	C1	
13.	company-method-name	string	C1	
14.	company-method-id	string	C1	
15.	pre-experiment-program-name	string	C5	
16.	post-experiment-program-name	string	C5	
17.	source-file-reference	string	C5	M5
18.	error-log	string	C5	

The Req'd column indicates whether a data element is required, and if required, for which categories. For example, M1234 indicates that that

AIA CHROMATOGRAPHY DATA STANDARD 189

particular data element is required for any dataset that includes information from category 1, 2,3, or 4; M4 indicates that a data element is only required for category 4 datasets.

Unless otherwise specified, data elements are generally recorded to be their actual test values, instead of the nominal values that were used at the initiation of a test.

3.1.1 DATASET-COMPLETENESS

Indicates which Analytical Information Categories are contained in the dataset. The string should exactly list the Category values, as appropriate, as one or more of the following "Cl + C2 + C3 + C4 + C5", in a string separated by plus(" + ')signs. This data element is used to check for completeness of the analytical dataset being transferred.

3.1.2 AIA-TEMPLATE-REVISION

The revision level of the AIA template being used by implementors. This needs to be included to tell users which revision of the **Specification** document should be referenced for the exact definitions of terms and data elements used in a particular dataset.

3.1.3 NETCDF-REVISION

The current revision level of the netCDF data interchange system software being used by the AIA for data transfer.

3.1.4 LANGUAGES

An optional list of natural (human) languages and programming languages delineated for processing by language tools.

1) ISO-639-language—indicates a language symbol and country code from Annex B and D of the ISO-639 Standard.

2) other-language—indicates the language and dialect using a user-readable name; only for those languages and dialects not covered by ISO 639 (such as programming language).

3.1.5 ADMINISTRATIVE-COMMENTS

Comments about the dataset identification of the experiment. This free text field is for anything in this information class that is not covered by the other data elements in this class.

3.1.6 DATASET-ORIGIN

The name of the organization, address, telephone number, electronic mail nodes, and names of individual contributors, including operator(s), and any other information as appropriate. This is where the dataset originated.

3.1.7 DATASET-OWNER

The name of the owner of a proprietary dataset. The person or organization named here is responsible for this field's accuracy. Copyrighted data should be indicated here.

3.1.8 DATASET-DATE-TIME-STAMP

Indicates the absolute time of dataset creation relative to Greenwich Mean Time. Expressed as the synthetic datetime given in the form: YYYYMMDDhhmmss+/-ffff.

This is a synthesis of ISO 2014 "Writing of Calendar Dates in All-Numeric Form', ISO 3307 "Information Interchange—Representations of Time of the Day', and ISO 4031 "Information Interchange—Representations of Local Time Differentials', which compensates for Local Time Variations.

The time differential factor (fffo expresses the hours and minutes between local time and the Coordinated Universal Time (UTC or Greenwich Mean Time, as disseminated by time signals), as defined in ISO 3307. The time differential factor (ffff) is represented by a four-digit number preceded by a plus (+) or a minus (-) sign, indicating the number of hour and minutes that local time differs from the UTC. Local times vary throughout the world from UTC by as much -1200 hours (west of the Greenwich Meridian) and by as much as +1300 hours (east of the Greenwich Meridian). When the time differential factor equals zero, this indicates a zero hour, zero minute, and zero second difference from Greenwich Mean Time.

An example of a value for this data element would be: 1991,08,01,12:30:23-0500 or 19910801123023-0500. In human terms this is 12:30 PM on August 1, 1991 in New York City. Note that the -0500 hours is 5 full hours time behind Greenwich Mean Time. 'Me ISO standards permit the use of separators as shown, if they are required to facilitate human understanding. However, separators are not required and consequently shall not be used to separate date and time for interchange among data processing systems.

The numerical value for the month of the year is used, because this eliminates problems with the different month abbreviations used in different human languages.

3.1.9 INJECTION-DATE-TIME-STAMP

Indicates the absolute time of sample injection relative to Greenwich Mean

AIA CHROMATOGRAPHY DATA 191

Time. Expressed as the synthetic datetime given in the form: YYYYMMDDhhmmss+/-ffff. See dataset-date-timestamp for details of the ISO standard definition of a date time stamp.

3.1.10 EXPERIMENT-TITLE

An optional user-readable, meaningful name for the experiment or test that is given by the scientist.

3.1.11 OPERATOR-NAME

The name of the person who ran the experiment or test that generated the current dataset.

3.1.12 SEPARATION-EXPERIMENT-TYPE

The name of the separation experiment type. Select one of the types shown in the following list. The full name should be spelled out, rather than just referencing the number. This requirement is to increase the readability of the datasets.

Users are advised to be as specific as possible, although for simplicity, users should at least put "gas chromatography" for GC or "liquid chromatography" for LC to differentiate between these two most commonly used techniques.

Separation Experiment Types

Gas Chromatography
1) Gas Liquid Chromatography
2) Gas Solid Chromatography

Liquid Chromatography
3) Normal Phase Liquid Chromatography
4) Reversed Phase Liquid Chromatography
5) Ion Exchange Liquid Chromatography
6) Size Exclusion Liquid Chromatography
7) Ion Pair Liquid Chromatography
8) Other

Other Chromatography
9) Supercritical Fluid Chromatography
10) Thin Layer Chromatography
11) Field Flow Fractionation
12) Capillary Zone Electrophoresis

3.1.14 COMPANY-METHOD-ID

The internal method id of the sample analysis method used by the company.

3.1.13 COMPANY-METHOD-NAME

The internal method name of the sample analysis method used by the company.

3.1.15 PRE-TEST-PROGRAM-NAME

The name of the program or subroutine that is run before the analytical test is finished.

3.1.16 POST-TEST-PROGRAM-NAME

The name of the program or subroutine that is run after the analytical test is finished.

3.1.17 SOURCE-RILE-REFERENCE

Adequate information to locate the original dataset. This information makes the dataset self referential for easier viewing and provides internal documentation for GLP-compliant systems.

This data element should include the complete filename, including node name of the computer system. For UNIX this should include the full path name. For VAX/VMS this should include the node-name, device-name, directory-name, and file-name. The version number of the file (if applicable) should also be included. For personal computer networks this needs to be the server name and directory path.

If the source file was a library file, this data element should contain the library name and serial number of the dataset.

3.1.18 ERROR-LOG

Information that serves as a log for failures of any type, such as instrument control, data acquisition, data processing or others.

3.2 Sample-Description Information Class

The SAMPLE-DESCRIPTION information class is comprised of nominal information about the sample. In the future this class will also need to contain much more chemical method and Good Laboratory Practice information.

SAMPLE-DESCRIPTION INFORMATION CLASS

	Data Element Name	Datatype	Category
1.	sample-id-comments	string	C5
2.	sample-id	string	C1
3.	sample-name	string	C1
4.	sample-type	string	C1
5.	sample-injection-volume	floating-point	C3
6.	sample-amount	floating-point	C3

3.2.1 SAMPLE-ID

The user-assigned identifier of the sample.

3.2.2 SAMPLE-ID-COMMENTS

Additional comments about the sample identification information that are not specified by any other SAMPLE-DESCRIPTION data elements.

3.2.3 SAMPLE-NAME

The user-assigned name of the sample.

3.2.4 SAMPLE-TYPE

Indicates whether the sample is a standard, unknown, control, or blank.

3.2.5 SAMPLE-INJECTION-VOLUME

The volume of sample injected, with a unit of microliters.

3.2.6 SAMPLE-AMOUNT

The sample amount used to prepare the test material. The unit is milligrams.

3.3 Detection-Method Information Class

This information class holds the information needed to set up the detection system for an experiment. Data element names assume a multichannel system. The first AIA implementation applies to a single-channel system only. This table shows only the column headers for a detection method for a single sample.

DETECTION-METHOD INFORMATION CLASS

	Data Element Name	Datatype	Category	Required
1.	detection-method-table-name	string	C1	
2.	detection-method-comments	string	C1	
3.	detection-method-name	string	C1	
4.	detector-name	string	C1	
5.	detector-maximum-value	floating-point	C1	m1
6.	detector-minimum-value	floating-point	C1	m1
7.	detector-unit	string	C1	m1

3.3.1 DETECTION-METHOD-TABLE-NAME

The name of this detection method table. This name is global to this table. It is included for reference by the Sequence information table and other tables.

3.3.2 DETECTION-METHOD-COMMENTS

The user's comments about detector method that is not contained in any other data element.

3.3.3 DETECTION-METHOD-NAME

The name of the detection-method actually used. This name is included for archiving and retrieval purposes.

3.3.4 DETECTOR-NAME

The user-assigned name of the detector used for this method. This should include a description of the detector type, and the manufacturer's model number. This information is needed along with the channel name in order to track data acquisition. For a single-channel system, channel-name is preferred to the detector-name, and should be used in this data element.

3.3.5 DETECTOR-MAXIMUM-VALUE

The maximum output value of the detector as transformed by the analog-to-digital converter, given in detector-unit. In other words, it is the maximum possible raw data value (which is not necessarily the actual maximum value in the raw data array). It is required for scaling data from the sending system to the receiving system.

3.3.6 DETECTOR-MINIMUM-VALUE

The minimum output value of the detector as transformed by the analog-to-digital converter, given in detector-unit. In other words, it is the minimum possible raw data value (which is not necessarily the actual minimum value in the raw data array). It is required for scaling data to the receiving system.

AIA CHROMATOGRAPHY DATA   195

3.3.7 DETECTOR-UNIT

The unit of the raw data. Units may be different for each of the detectors in a multichannel, multiple detector system.

NOTE **ABOUT DATA SCALING:** Data arrays are accompanied by the maximum and minimum values (detector—maximum, detector—minimum, and detector—unit) that are possible. These can be used to scale values and units from one system into values and units for another system. For example, one system may produce raw data from 0 to 100000 counts, and be converted to -100 millivolts to 1.024 volts on another system. This scaling is not done automatically, and must be done by either the sending or receiving system if required.

3.4 Raw-Data Information Class

This is the information actually generated by the data acquisition process. The data are then fed into the peak processing algorithms. This table shows only the column headers for the raw data arrays. Figure 1 illustrates the exact meaning of the data elements in this information class.

RAW-DATA INFORMATION CLASS

Data Element Name	Datatype	Required
1. point-number	dimension	M1
2. raw-data-table-name	string	
3. retention-unit	string	M12
4. actual-run-time-length	floating-point	M12
5. actual-sampling-interval	floating-point	M12
6. actual-delay-time	floating-point	M12
7. ordinate-values	float-array	M1
8. uniform-sampling-flag	Boolean	M1
9. autosampler-position	string	
10. raw-data-retention	float-array	M1, is uniform-sampling-flag N'

3.4.1 POINT-NUMBER

The value of point-number is the dimension of the ordinate-values and (if present) raw-data retention arrays. It should be set to zero if these arrays are empty.

3.4.2 RAW-DATA-TABLE-NAME

The name of this table, included for reference by the Sequence information table and other tables.

3.4.3 RETENTION-UNIT

The unit along the chemical or physical separation dimension axis. All other data elements that reference the separation axis have this same unit.

NOTE: The AIA Committee has considered the implications and relative merits of using time versus volume, and is using a 'seconds" unit for chromatography techniques, including Capillary Zone Electrophoresis (CZE) and Size Exclusion Chromatography (SEC). If the user employs CZE or SEC, and wants to use a unit other than seconds, then he should use that unit as the value of the retention-unit data element.

For liquid and gas chromatography the default unit for the retention axis is time in seconds.

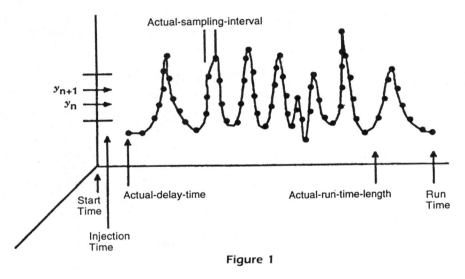

Figure 1

3.4.4 ACTUAL-RUN-TIME-LENGTH

The actual run time length from start to finish for this raw data array, given in the retention-unit.

3.4.5 ACTUAL-SAMPLING-INTERVAL

The actual sampling interval used for this run, given in the unit of the retention-unit. At this time, it is for a fixed sampling interval.

3.4.6 ACTUAL-DELAY-TIME

The time delay between the injection and the start of data acquisition, given in the retention-unit.

AIA CHROMATOGRAPHY DATA

3.4.7 ORDINATE-VALUES

This is a set of values of dimension **point-number,** containing the ordinate values. This set of values has a unit of **detector-unit.** This is a required field for datasets containing raw data.

There is no data point at time = 0. 0 (or volume = 0. 0). The first data point is at the first point after the start of data acquisition.

3.4.8 UNIFORM-SAMPLING-FLAG

A value of 'N" for this flag indicates that some kind of non-uniform sampling was used. If non uniform sampling was used, then an array for raw data retention is required. The default value for this is "Y".

3.4.9 AUTOSAMPLER-POSITION

The position in the autosampler tray. The default datatype for this was chosen to be a string because some companies have concentric rows of sample vials in the sample tray; others may use Cartesian coordinates.

The format of this is a free-form string, with two sub-strings, using a period as a delimiter, e.g., "coordinate1.coordinate2' or (tray.vial).

3.4.10 RAW-DATA-RETENTION

This is a set of values of dimension point-number, containing the abscissa value for each raw data ordinate value. This set of values has a unit of retention unit. This is a required field if the uniform sampling flag is "N".

Example:

raw-data (I) = 12   raw-data-retention (1) = 0.2

raw-data (n-1) = 998760   raw-data-retention (n-1) = 120.1
raw-data (n) = 997650   raw-data-retention (n) = 120.3

raw-data (n + 1) = 996320   raw-data-retention (n + 1) = 121.5

raw-data (point_number) = 20   raw-data_retention (point_number) = 720.2

NOTES FOR USAGE OF RAW-DATA INFORMATION:
The order of usage for using raw data from this information class is very simple. First check the **uniform-sampling-flag** to see if it is "Y". If it is, then

use only the ordinate-values array for amplitude values, and calculate the abscissa values from point 0.0 onward using the **actual-sampling interval**. If the value of **uniform-sampling-flag** is "N", then use the ordinate-values array for amplitude values and the raw data retention array for abscissa values.

3.5 Peak-Processing-Results Information Class

This is the information generated by the peak processing algorithms. *Final processed results may vary from manufacturer to manufacturer.*
PEAK-PROCESSING-RESULTS Information Class

	Data Element Name	Datatype	Category	Req'd
1.	peak-number	dimension	C2	M2
2.	peak-processing-results-table-name	string	C3	
3.	peak-processing-results-comments	string	C2	
4.	peak-processing-method-name	string	C2	
5.	peak-processing-date-time-stamp	string	C2	
6.	peak-retention-time	floating-point-array	C2	M2
7.	peak-name	string-array	C3	
8.	peak-amount	floating-point-array	C2	M3
9.	peak-amount-unit	string	C2	M3
10.	peak-start-time	floating-point-array	C2	
11.	peak-end-time	floating-point-array	C2	
12.	peak-width	floating-point-array	C2	
13.	peak-area	floating-point-array	C2	M2
14.	peak-area-percent	floating-point-array	C2	
15.	peak-height	floating-point-array	C2	M2
16.	peak-height-percent	floating-point-array	C2	
17.	baseline-start-time	floating-point-array	C2	
18.	baseline-start-value	floating-point-array	C2	
19.	baseline-stop-time	floating-point-array	C2	
20.	baseline-stop-value	floating-point-array	C2	
21.	peak-start-detection-code	string-array	C2	
22.	peak-stop-detection-code	string-array	C2	
23.	retention-index	floating-point-array	C2	
24.	migration-time	floating-point-array	C2	
25.	peak-asymmetry	floating-point-array	C2	
26.	peak-efficiency	floating-point-array	C2	
27.	mass-on-column	floating-point-array	C2	
28.	manually-reintegrated-peaks	boolean-array	C2	

3.5.1 PEAK-NUMBER

The peak number used to identify a particular peak. The value of peak-number is the dimension of each array in the peak-processing-results information class. This data element should be set to zero if no peak processing results are included in the dataset.

3.5.2 PEAK-PROCESSING-RESULTS-TABLE-NAME

The name of this table, included to make the dataset self-referential. It is global to this information class.

AIA CHROMATOGRAPHY DATA

3.5.3 PEAK-PROCESSING-RESULTS-COMMENTS

Comments about the peak processing results that are not contained in any other data element in this information class.

3.5.4 PEAK-PROCESSING-METHOD-NAME

The name of the method used for peak processing. This is typically assigned by the end user. It is typically used for archival and retrieval purposes.

3.5.5 PEAK-PROCESSING-DATE-TIME-STAMP

The date time stamp for peak processing. Indicates when the data was processed. See **datasetdate-time-stamp** for ISO standard date time stamp syntax details.

3.5.6 PEAK-RETENTION-TIME

The retention time of the peak detected, given in the unit of retention-unit.

3.5.7 PEAK-NAME

The user-assigned name of the peak. This is an optional field because some peaks may be unknown.

3.5.8 PEAK-AMOUNT

The amount of substance that is determined by peak processing.

3.5.9 PEAK-AMOUNT-UNIT

The unit used for the peak amount. This is in a concentration unit, absolute amount in grams, or some other appropriate unit.

3.5.10 PEAK-START-TIME

The starting point of the peak, given in a unit of retention-unit.

3.5.11 PEAK-END-TIME

The ending point of the peak, given in a unit of retention-unit.

3.5.12 PEAK-WIDTH

The calculated width of the peak, given in a unit of **retention-unit.**

3.5.13 PEAK-AREA

The computed area of the peak, given in a scaled data unit of **(detection-unit * retention-unit).**

3.5.14 PEAK-AREA-PERCENT

The computed area percent of the peak: the summation of all quantified peaks in an analysis should equal 100.0.

3.5.15 PEAK-HEIGHT

The computed height of the peak; in a scaled data unit. The unit for this is the same as that of the ordinate-values.

3.5.16 PEAK-HEIGHT-PERCENT

The computed height percent of the peak; the summation of all quantified peaks in an analysis should equal 100.0.

3.5.17 BASELINE-START-TIME

The starting point of the computed baseline for this peak, given in a unit of **retention-unit.**

3.5.18 BASELINE-START-VALUE

The starting value of the computed baseline for this peak; in a scaled data unit, the unit for this is the same as that of the ordinate variable.

3.5.19 BASELINE-STOP-TIME

The ending point of the computed baseline for this peak, given in a unit of **retention-unit.**

3.5.20 BASELINE-STOP-VALUE

The starting value of the computed baseline for this peak; in a scaled data unit, the unit for this is the same as that of the ordinate variable.

3.5.21 PEAK-START-DETECTION-CODE

Codes that are used to describe how the baselines have actually been drawn. The peak type may be represented by a two-letter code.

The following are examples of peak detection codes:

1. B-baseline peak, i.e., the peak begins and/or ends at the baseline.
2. P-perpendicular drop, i.e., the peak begins and/or ends with a perpendicular drop.
3. S-skimmed peak, i.e., the peak is a shoulder peak that is skimmed.
4. VD-vertical drop, i.e., the peak begins or ends at a vertical drop to the skim line (such as between two skimmed peaks).

AIA CHROMATOGRAPHY DATA 201

5. HP-horizontal projection, i.e., the peak baseline starts and/or ends with a horizontal projection.

6. EX-exponential skim, i.e., the peak starts and/or stops with an exponential skim.

7. PT-pretangent skim, i.e., the leading edge of the peak is tangent skimmed.

8. MN—manual peak, i.e., the user forced the baseline at the data level.

9. FR—forced peak, i.e., the user forced the baseline at a user-supplied level.

10. DF—user forced daughter peak, i.e., the user forced a baseline on the side of a fused peak (daughter peak).

11. LP—lumped peak, i.e., peaks values are lumped (added) together until this timed event is turned off.

3.5.22 PEAK-STOP-DETECTION-CODE

See **peak-start-detection-code** for definition.

3.5.23 RETENTION-INDEX

The retention index, as defined by Kovats, is a measure of relative retention which uses the normal alkanes as a standard reference. Each normal hydrocarbon is assigned a number equal to its carbon number times one hundred. For example, n-pentane and n-decane are assigned indices of 500 and 1000, respectively. Indices are calculated for all other compounds by logarithmic interpolation of adjusted retention times, as shown in the following equation.

$$Ia = 100^*N + 100^*n^* \left[\frac{\log tRa - \log tRN}{\log tR(N+n) - \log tRN} \right]$$

where

Ia	is the retention index of the peak
N	is the carbon number of the lower n-alkane
·	is the difference in carbon number of the two n-alkanes that bracket the compound
tRa	is the adjusted retention time of the unknown compound
tRN	is the adjusted retention time of the earlier eluting n-alkane that brackets the unknown
tR(N + n)	is the adjusted retention time of the later eluting n-alkane that brackets the unknown

3.5.24 MIGRATION-TIME

The transit time from the point of injection to the point of detection, given in the retention-unit. This is used in Capillary Zone Electrophoresis (CZE).

3.5.25 PEAK-ASYMMETRY

The peak asymmetry measured as $A, = B/A$, where A and B are the widths for the front and back parts of a peak, commonly measured at 10% peak height for USP at 5%.

3.5.26 PEAK-EFFICIENCY

Also known as the column theoretical plate number for an individual peak. The peak-efficiency can
be expressed as

retention-time / width-at-half-height 2] $\times$ 5.54 or as
retention-time / baseline-bandwidth 21 $\times$ 16

where baseline-bandwidth is the peak width along the baseline as determined by the intersection points of the tangents drawn to the peak above its points of inflexion. This is usually measured for a reference component, although routinely a representative peak in the chromatogram is chosen and this is reported for it.

3.5.27 MASS-ON-COLUMN

A measure of column loading. It is usually reported as the sum of the peak-amount(s). It needs to be determined against known peaks.

3.5.28 MANUALLY-RE INTEGRATED-PEAKS

A Boolean flag that indicates if any reported results are based on manual manipulation of baselines and/or peak start/end times. A value of logical "O" for this flag indicates that the current peak was not manually re integrated.

APPENDIX A-REFERENCES

1. **AIA Chromatography Data Standard Implementation Guide,** Analytical Instrument Association—Committee on Communications Standard. Copies can be obtained from the Analytical Instrument Association, 225 Reinekers Lane, Suite 625, Alexandria, VA 22314.

2. **netCDF User's Guide,** by Russell **K.** Rew, Unidata Program Center, University Corporation for Atmospheric Research, P.O. Box 3000, Boulder, CO 80307-3000

3. ISO Standard 2014-1976 (E) **Writing of Calendar Dates in AU-Numeric Form**

4. ISO Standard 3307-1975 (E) **Information Interchange—Representations of Time of the Day**

5. ISO Standard 4031-1978 (E) **Information Interchange—Representations of Local Time Differentials**

Index

Access control:
  networking and, 82–83
  system security and, 93–94
Administrative information class, AIA data format standard, 188–189
Administrative servers, client-server systems, 158
Adobe Photoshop system, 159
Algorithms:
  AIA data format standard, 184
  post-run data analysis, 28–29
American Society for Testing Materials (ASTM):
  LIMS Guide, 35
  system development procedures, 98
Analog-to-digital converter (A/D), serial communications, 170–172
Analytical information categories, AIA data format standard, 185–188
Analytical Instrumentation Association (AIA):
  Chromatography Data Standard Specification, 121
  administrative information class, 188–192
  analytical information categories, 186–188
  applications, 121
  detection-method information class, 193–195
  full text of, 177–201
  implementation strategy, 185
  raw-data information class, 195–201
  sample-description information class, 192–193
  technical objectives, 183–185
  standards research at, 123–127
Asset management:
  information as asset, 48–49
  laboratory notebook/diary example of, 49–52
  productivity and, 53–54
  technology planning and implementation, 48–52
Auto-injector systems, 111
  historical overview, 121–122
Automated Systems Integration Corporation, 109–110
Automatic samplers, laboratory robotics, 111

Backbone, of network systems, 173–176
Backup:
  data librarian database design, 130
  electronic laboratory notebooks (ELNB), 138–139
  of information, 50–51
  network systems design and, 82–83, 172–173
  system procedures for, 97
  types of, 52
Barcode scanners, automated sample storage, 109–110
Bitfile Mover, hierarchical storage systems, 153
Bitfile server, hierarchical storage systems, 153

Call-backs, system security with, 95
Carpal tunnel syndrome, 44
Catalogs:
  of information, 51
  searching, in data librarians, 131
Central processing unit (CPU):
  farms, 161–163
  power capabilities, 159
Change agent, computing technology as, 38–39

INDEX

Checksums mechanisms, laboratory robotics, 115–116
Chromatography:
  cost/value analysis of, 53–54
  data format:
    AIA standard specification, full text, 177–201
    application of standard, 121
    as data acquisition systems model, 32–34
    historical overview of technology, 121–123
    laboratory automation models, 119, 122
    netCDF data automation and, 126–127
Client-server systems:
  electronic laboratory notebooks (ELNB), 139
  laboratory automation with, 156–159
  operating system design, 146–150
Common database, information management through, 66–67
"Common knowledge" oval, service laboratories and, 72–73
Communications protocol:
  laboratory robotics and computer system, 115–116
  LASF model, 24–25
  network systems, 169–176
  serial communications, 170–172
Compression techniques, mass storage management, 155–156
Computer viruses:
  protection packages, 97
  system security and, 95–96
Computing capability, *see* Power capability
Computing technology:
  current state of, 4–8
  historical background on, 8–12
  laboratory applications:
    client-server systems, 156–159
    data storage, 150–156
    hyper-information systems, 164–169
    networks and communications, 169–176
    operation systems, 143–150
    overview, 142–143
    power considerations, 159–164
    robotics linked with, 114–116
  needs pyramid for, 10–12
  planning and implementation guidelines, 46–59
    asset management, 48–52
    corporate strategies, 47
    goal setting, 47–48
    information flow, 58–59
    integration strategies, 54–58
    productivity as asset, 53–54
  strategies regarding:
    human factors in, 38–46
    implementation concerns, 46–59
    planned approach, 37–38
  use patterns and problems, 6–8
Conferencing technology, electronic laboratory notebooks (ELNB), 141
Consortium for Automated Analytical Laboratory Systems (CAALS):
  CAALS-1 Communications Specification, 124
  standards research, 123–124
Control systems:
  instruments and measuring devices, 119, 121–123
  laboratory robotics, 114–116
Conversion of information, 56–57
Cost/value analysis:
  impact on computing technology, 7–8
  laboratory robotics, 112–114
  make or buy decisions, 88–90
  paper assets, 49–52
  productivity and, 53–54

Data, *see also* K.I.D. diagram; Storage of data acquisition systems:
  K.I.D. diagram, role in, 55–58
  laboratory automation, history, 121–123
  modularity and, 30–35
  AIA data format standards, 124–127
    applications, 124–127
    full text of standard, 177–201
    computed results format, 126
    consistency of, need for, 27–28
    defined, 54
    formatting and organization of, 67–68
    full chemical method, 126
    full data processing model, 126
    GLP information format, 126
    integrity across heterogeneous systems, AIA standard, 183–184
    LASF model, 24
    raw data category, 126
    AIA absolute scaling, 184
    AIA analytical information categories, 186, 195–196
    sample layouts for, 99–100
Data Interchange System (DIS), AIA data format standard, 185
Data Librarian:
  database design, 129–130
  file servers and, 150
Data librarian:
  implementation strategies, 129–132
  laboratory automation, 127–129

INDEX

Linda Technology and, 164
list of functions, 131
role of, in LASF model, 27-28
scalability and growth paths, 129
security concerns, 130
simplicity of design, 131-132
Database design, data librarians, 129-130
DECnet system, 84
Dedicated boxes, data acquisition systems, 31-32
Deskillinization, technology strategies and, 39-41
Detection-method information category, 193-194
Disaster recovery, off-site information storage, 51
Disk compression utility, operating system design, 148
Disk operating system (DOS), historical overview, 145-146
DiskDouble system, operating system design and, 149
Distributed batch processing, power capabilities and, 160-161
Distributed processing:
file servers and, 151
power capabilities and, 159-160
Dynamic Data Exchange software, 140

EIA-485 system, laboratory automation technology, 124
Electromagnetic fields, computing technology strategies and, 43-45
Electronic mail (EMAIL):
hardware requirements for, 80
human attitudes toward, 41-43
intergroup communication through, 79-81
storage requirements, 80
Electronics laboratory notebook (ELNB), 137-141
hyper-information systems, 167-168
Environmental issues, computing technology and, 43-45
Evaluation, LASF model, 24-25

Falling letters virus, 96
FAX, intergroup communication through, 79-81
FDA-483 citations, implementation planning, 91-92

File processing, in data librarians, 131
File servers, data storage with, 150-151
Fourier Transform Infrared Spectrometry Standard, 127
Fragmentation, system maintenance and, 99-101

Goal setting, technology planning and implementation, 47-48
Good Automated Laboratory Automation Guidelines, 92
Good Automated Laboratory Practices (GALP), 76-77
Good Laboratory Practices (GLP):
AIA data format standard and, 185, 187
data presentation, 126
Graphics programs, lack of standardization in, 70-71
Groupware:
development of, 69-70
electronic laboratory notebooks (ELNB), 140-141
Growth paths, data librarians, 129

Hardware:
changes in specifications, 13-14
development procedures for, 97-99
fragility of market dominance with, 123
hierarchical storage and, 154-156
maintenance procedures, 99-101
needs pyramid for, 10-12
operating system design and, 149-150
operating systems, historical overview, 144-145
product evaluation, 84-86
projections of future requirements, 4, 9
software design independent of, 69-70
Health issues, strategic planning and, 43-45
Hierarchical storage systems, 152-156
"Hot links," software upgrading with, 85-86
HP-IB interface, laboratory robotics, 118-119
Human factors:
computing technology strategies and, 38-46
electronic mail, 41-43
environmental concerns, 43-45
organizational structure, 41
voice input and output, 45-46
work environment considerations, 38-41
electronic laboratory notebooks (ELNB), 141
make or buy decisions, 89-90

INDEX

Hyper-information systems:
 electronic laboratory notebooks (ELNB), 138–139
 laboratory automation with, 164–169
HyperCard system, 165
Hyperslab function, netCDF data automation, 126
Hypertext, 165–166

I/O functions:
 laboratory robotics, 118–119
 serial communications, 170–172
IBM:
 as catalyst in personal computer market, 10–12
IBM–PC:
 historical overview of, 122–123
 instruments and measuring devices, 119
IEEE-Mass Storage System Reference Model (IEEE-MSSRM), 152–156
 data librarian database design, 129–130
IEEE-488 bus, laboratory automation technology, 124
Image enhancement, CPU farms, 162
Implementation plan for laboratory automation:
 development of, 62–64
 electronic mail, role of, 79–81
 FDAS-483 citations, 91–92
 K.I.D. diagram information flow, 64–77
  intergroup communication, 77–84
  service laboratories, 72–77
 LIMS automation strategies, 134–135
 make or buy decisions, 88–90
 network considerations, 81–84
 overview, 47
 product and technology evaluations, 84–86
 software revisions and updates, 86–87
 standard operating procedures, 92–102
  backup and recovery, 97
  maintenance, 99–101
  software and systems development, 97–99
  system security, 93–97
  validation, 101–103
 technology overview, 107–108
 testing procedures, 90–91
Information strategies. *see also* K.I.D. diagram
 alternative service laboratory flows, 74–75
 as asset, 48–49
 computing technology strategies and, 37
 defined, 54
 flow of, in K.I.D. diagram, 58–59
 in FAX form, 80–81
 laboratory robotics, 112–116
 LASF model, 24

service laboratories and, 72–77
vendor alliances in laboratory automation and, 121–122
Infrared spectroscopy standard, future development of, 121
Instrumentation:
 effects of emissions on, 43–44
 historical overview, 121–123
 laboratory automation models, 119–123
 LASF model, 24–25
 standardization of data formats for, 75–77
Integration of systems:
 defined, 57
 future need for, 12
 information's role in, 54–58
 modularity and, 29
 product evaluation for capability of, 85–86
Intergroup communication, implementation planning and, 77–84
International Symposium on Laboratory Automation and Robotics (ISLAR), 112–114
ISO 9000 regulations:
 AIA data format standard and, 185, 187
 computing technology strategies and, 36, 46
Isolation:
 avoidance of, with LASF model, 25
 EMAIL as factor in, 42–43

Junk email, 42
Justification strategies, computing technology planning with, 36–37

K.I.D. (knowledge/information/data) model:
 "Common knowledge" oval, 72–75
 implementation planning with, 63–64
 movement of information through, 64–77
 information exchange in, 58–59
 integration of information in, 54–58
 lab process models and, 76–77
 LIMS automation strategies and, 133–135
Keyboards, environmental issues regarding, 44–45
Knowledge, *see also* K.I.D. diagram
 defined, 54
 LASF model, 24

Laboratory automation models:
 AIA data format standard, 124–127
 data acquisition systems, modularity and, 30–35
 data librarian, 27–28, 127–129

INDEX

electronic laboratory notebooks, 137–141
historical background, 19–21
instruments and measuring devices, 119–123
Laboratory Information Management System (LIMS), 132–137
LASF model, 21–35
   communications, 25
   instruments and evaluators, 24–25
   knowledge, information and data management and use, 24
   material handling and management, 21, 24
   modularity, 29–30
   NIST/CAALS protocol, 123–124
   overview, 18–19
   post-run data analysis, 28–29
   sample logins, 25–26
   testing and scheduling, 26
Laboratory Automation Standards Foundation (LASF):
   background, 19
   communications, 25
   data librarian, 128–129
   instruments and evaluators, 24–25
   knowledge, information and data, 24
   material handling and management, 21, 24
   schematic, 22–23
Laboratory Information Management System (LIMS):
   automated sample storage in, 111
   automation strategies, 132–137
   client-server systems, 157
   concept functions, 32–34
   data librarian, 128–129
   failure in, causes of, 136–137
   file servers and, 151
   historical background, 133
   implementation planning for, 62–63
   laboratory automation models, 21
   laboratory workload considerations, 135–137
   networking with, 81–84
   pros and cons of, 136
Laboratory management system (LMS):
   automated sample storage, 110–111
   file servers and, 151
Languages, AIA data format standard, 189
Lighting, strategic planning and, 44
Linda Technology, power capabilities and, 163–164
Local Area Networks (LANs), human factors in, 39–40
Local Area Transport (LAT), serial communications, 170–172

LocalTalk network, 81–84
Loyalty to programs, human factors in development of technology and, 40–41

Macintosh:
   communications protocol, 169–170
   as example of integrated user interface, 12
   voice input and output with, 45
Magnetic Field Menace, 43
Maintenance functions:
   in data librarians, 131
   procedures for, 99–101
Make or buy decisions:
   implementation strategies, 88–90
Market dominance:
   as driving force for laboratory technology, 123
Marketplace mentality:
   operating system design and, 148–149
Maslow's Pyramid of Human Needs, 9–10
Mass spectrometry:
   historical overview, 121–122
   standard for, future development of, 121, 127
Material handling and management:
   LASF model, 21, 24
Measuring devices, *see also* specific devices
   laboratory automation models, 119–123
Media care and management, 51
MEMEX machine, 165
Memory capacity, operating systems, 145
"Methods" manager, in data librarian, 131
Microwave radiation, computing technology strategies and, 43–45
Modularity:
   data acquisition systems and, 30–35, 146–147
   in LASF models, 29–30
   operating system design, 147–150
MS-DOS:
   as industry standard, 11–12
   data formatting and organization on, 68
   updating strategies in, 86–87
Multi-task operating systems:
   historical overview, 146
   laboratory automation applications, 143–144
   work environments, 84
Multiprocessor systems, CPU farms, 162–163

Name Server, hierarchical storage systems, 153

INDEX

National Institute for Standards and Technology (NIST), 116
National Research and Education Network (NREN):
  applications, by bandwidth and traffic, 8
  overview of, 4
  timetable for, 7
"Near-line" backup systems, 130
  hierarchical storage and, 153–156
Nerve and joint ailments, keyboard use, 44–45
netCDF (Network Common Data Form), 125–127
  AIA data format standard:
    information categories, 185–186
    revision, 189
Network systems:
  client-server systems, 157
  communications protocol, 169–176
  components of, 173–176
  design and management considerations, 172–173
  hierarchical storage in, 153–156
  hyper-information systems, 167–168
  implementation planning of, 67–68, 81–84
  limits of, 82–83
  system security and, 96–97
NIST/CAALS communications protocol:
  laboratory robotics and computer systems, 116
  overview, 123–124
  serial communications, 170–172
Noise levels, role of, in strategic planning, 44–45
NSFNET, as computing model, 4–6

Off-site information storage, importance of, 51
Operating systems:
  data acquisition function, role of, 146–149
  data formatting and organization problems and, 68
  impact on software development, 12
  laboratory automation applications, 143–150
    historical overview, 144–150
    multi-task systems, 143–144
    single task systems, 143
    time-sharing systems, 144
  modifiers in, 31
  modularity and, 30–35
  product evaluation, 84–86
  standardization of:
    need for, 11–12

Orca product line (Hewlett-Packard):
  integration with labor instruments, 120
  laboratory robotics, 111–116
  technology overview, 116–119
Organizational structure, in technology strategies, 41
Out-board data acquisition systems, 31–32
Out-of-specification results, robotics systems, 115–116

Page layout software, electronic laboratory notebooks (ELNB), 139–140
Pagemaker software, 139–140
Paper information:
  cost/value relationship of, 49–52
  distribution issues, 53–54
Parallel processing:
  CPU farms, 161–162
  power capabilities and, 159
Passwords, use of, 94–95
Pathworks, file server design, 150
Peak-processing results information, 195–202
Personal computers, development of, 10–12
Photocopying, cost-value relationships of, 52
Physics applications, CPU farms, 162–163
Planning strategies, overview, 3–4
Policy making, information as asset in, 51–52
Post-run data analysis, in LASF model, 28–29
Power capability:
  laboratory automation, 159–164
    CPU farms, 161–163
    distributed batch processing, 160–161
    distributed processing, 159–160
    Linda technology, 163–164
  recent developments in, 14–15
Printing servers, client-server systems, 158
Priorities, email as catalyst in, 42
Process control:
  intergroup communication, 77–79
  K.I.D. diagram, role in, 55–58
  laboratory automation, 18–19
  login events, 25–26
  networking and, 81–84
  testing procedures, 26
Product evaluation, implementation planning and, 84–86
Product layering, operating system design and, 148–149
Productivity, asset management as, 53–54
Profitability, computing strategy and, 7–8

INDEX

Programming:
  laboratory robotics and, 114–116
  operating systems, 144–145
Proprietary serial protocols, 117–119
Pseudo-disk technology, operating system design, 148
Publish and Subscribe software, 140

Quality control:
  K.I.D. diagram, 56–57, 60
  laboratory automation, 19
QuickTime system, operating system design and, 149

Radiation, computing technology strategies and, 43–45
Real-time considerations, operating systems:
  design considerations, 147–149
  historical overview, 146
Recovery, system procedures for, 97
Redundant Array of Inexpensive Disks (RAID) systems:
  data librarian backup with, 130
  file servers and, 150–151
Research:
  management of, as asset, 49–52
  networking and, 83–84
Robotics systems:
  automated sample storage, 111
  data acquisition, 31–32
  equipment access as factor, 117–118
  laboratory robotics, 111–116
  networking of information and, 83–84
  stand-alone vs. computer-integrated systems, 114–116
  technology overview, 116–119
RS-232C interface:
  laboratory robotics, 118–119

Sample storage and preparation:
  AIA chromatography data format standard, 192–193
  automated model for, 108–111
  laboratory robotics, 111–116
Scalability:
  in data librarians, 129
  mass storage management, 155–156
  raw data, AIA standard, 184
Search strategies:
  hyper-information systems, 166–167
  in data librarian, 131–132

Security, see also Access
  in data librarians, 130
  implementation planning, 93–97
Self-esteem:
  email as threat to, 42–43
  technology strategies and, 39–41
Semiautonomous workstations, network systems, 173–176
Separation experiments, AIA chromatography data format standard, 191
Serial communications:
  laboratory automation, 169–176
  laboratory robotics with, 114
Servers, software updates run from, 87. See also Client-servers
Service laboratories, implementation planning with, 72–77
Single-task operating systems, laboratory automation applications, 143
Single-user systems, operating system technology, 146
Software:
  development procedures for, 97–99
  historical overview of laboratory technology and, 122–123
  human factors in development of, 40–41
  hyper-information systems, 165–169
  maintenance procedures, 99–101
  needs pyramid for, 11–12
  network system design and, 175–176
  operating systems impact on, 12
  product evaluation, 84–86
  revisions and updates, planning for, 86–87
  standardization of, independent of hardware, 69–77
  version numbers, strategies for using, 87
Spectrometry:
  cost/value analysis of, 53–54
  data format, data acquisition systems model, 32–34
  laboratory automation models, 119–122
  netCDF data automation and, 125–127
Standard operating procedures (SOPs), implementation planning, 92–102
  backup and recovery procedures, 97
  maintenance procedures, 99–101
  software and systems development, 97–99
  system security, 93–97
  validation procedures, 101–103
Standardization:
  AIA data format standards, 124–127
  implementation planning and, 68–77
  instrument data formats, 75–77
  instruments and measuring devices, 119, 121

Standardization (Continued)
  LASF models, 26–27
  NIST/CAALS standards, 122–124
  of operating systems, 11–12
  of software, 11
  of test procedures, 29
Storage of data:
  automated sample storage management, 108–111
  client-server systems, 156–159
  file servers, 150–151
  hierarchical systems, 152–156
  hyper-information systems, 168–169
  implementation planning and, 67–68
  K.I.D. diagram, role in, 55–58
  large-scale data storage, 155–156
  near-line backup system, 130
  technology implementation and, 150–156
StorageTek modules, hierarchical storage and, 153–156
Strawberry Tree Incorporated's Workbench MAC software, 13–15
Supervisory systems, distributed batch processing, 160–161
Support-person costs, evaluation of, 64
Synthesis, of information, 56–57

Tape librarian:
  automated sample storage, 111
  Dataserver, hierarchical storage and, 154–156
Testing periods, operating system design and, 148–149
Text handling, standardization of, 71–77
Time stamps, AIA data format standard, 190–191
Time-sharing systems:
  historical overview, 145–146
  laboratory automation applications, 144
  validity of, 64
Titration experiments, laboratory robotics with, 114
Tools, computer technology implementation and, 47
Training of workers, role of, in technology strategies, 39–41

Trojan horse virus, system security and, 95–96
Tupple and tupple spaces, Linda Technology, 163–164

UFMULTI system, CPU farms, 162–163
Unitree software:
  data librarian database design, 130
  hierarchical storage systems, 153, 155
University Corporation for Atmospheric Research, 125
Unix system:
  as industry standard, 11–12
  instruments and measuring devices, 119
Updating capability:
  planning strategies for, 86–87
  product evaluation for, 85–86
User-interface capability, 15–16

Validation:
  implementation planning with, 65–67
  modularity and, 29–30
  planning and implementing computer technology, 46
  standard procedures for, 101–103
Vendor alliances, laboratory automation technology and, 122–123
Video display terminals, 44
Visicalc, development of, 8–9
Voice input and output, strategic planning regarding, 45–46
Voice Navigator, voice input and output with, 45–46

Waters LACE box, 31
Windows, and laboratory robotics, 119
Wireless technologies, environmental issues in, 44
Work environment:
  environmental and health issues in, 43–45
  human factors in, 38–41
Worms, system security and, 95–96

Zymark Benchmate series:
  laboratory robotics, 111–116
  operating system design, 148
Zymark Z845 interface, 114
Zymark Zymate series, 116–119